Praise for Kathleen Dean Moore's *Riverwalking*

"Splendid, short riverside rustications . . . Moore's collection
sparkles as much as one of her sun-dappled streams."
— *Kirkus Reviews* (pointer)

"The essays in *Riverwalking* are as fluent as the rivers Kathleen
Dean Moore hikes beside, boats upon, and walks in. And
they're just as surprising. The meaning of happiness, the
rewards of poking around, the need for domesticity,
motherhood, death, or the peculiar reproductive habits of the
roughskin newt—anything might be waiting around the next
bend."   — Janet Lembke, author of *Skinny Dipping*

"In *Riverwalking*, Kathleen Dean Moore has managed to pull off
that hardest of literary balancing acts, blending an evocative
lyricism with precise observation, harnessing both in a moving
and informative tribute to the miracles found both in moving
waters and the hearts and souls of the men and women who
cherish riverine beauty as Moore does herself."
— W. D. Wetherell, author of *Upland Stream* and *Vermont River*

"Let us welcome a new voice in nature writing . . . a literary
companion as delightful to travel with as Robert Louis
Stevenson, Henry Thoreau or John Muir."
— *Tacoma News Tribune*

"It is a joy to walk rivers with Kathleen Dean Moore. Her prose is as clear and lucid as the bright waters she loves, and her responses, memories, and connections are as surprising and complex as the life she contemplates along a river's banks and under its slippery stones."

—Annick Smith, author of *Homestead*

"*Riverwalking* is a smart, compassionate, and wise meditation on living in place. Kathleen Moore creates currents of natural philosophy that move our spirit toward a sustainable vision."

—Terry Tempest Williams, author of *Refuge*

"Kathleen Dean Moore . . . is a master of the essay. Her style is open and relaxed. Deceptively so, really, because what she has to say is very often genuinely profound. . . . *Riverwalking* is powerful . . . a moving and vital little book. . . . Moore has found the connections among the land, the human race, life and the souls of everything around us. Reading this book is like being taken by the hand and guided through something very new."

—*Salem Statesman Journal*

"Sometimes fussy, even cantankerous, other times reverent, and at all times elegant and celebratory, these lovely essays trade the precise for the leisurely, then return to the precise, as gracefully as the movement of water. It is a delight to travel the length of this book."

—Rick Bass, author of *The Platte River*

"*Riverwalking* is more than a fine book of essays. It's important—it's pointing the way toward a new kind of nature writing, one where the outdoors is in dialogue not only with our inmost souls but with our families, our relationships, our lives. Something powerful is at work here."

—Bill McKibben, author of *The End of Nature*

" 'There is one thing you will need to learn,' a philosophy professor admonished [Moore] in college, 'philosophy is not about life. Philosophy is about ideas. Life and ideas are not the same.' Well, Moore may have paid dutiful respect to this authoritative statement, but her own book disproves that effort. . . . This is a meditation on doing and thinking, beginning with rivers."   —*Los Angeles Times Book Review*

"Moore . . . will likely provoke a reader's own contemplation. Her prose has the kind of clarity the sun provides between clouds, the queer amplification created by clear water moving over cobble, when each stone seems individually etched."

—*California Fly Fisher*

"Spare and focused, Moore's essays meander in and out of the heart's own territory, like poetry. . . . Readers can delight in the evidence that Moore is willing to weep, to brood, to laugh uproariously as the loon, 'yowling and exultant,' to reveal to us her vulnerability and her deep humanity."   —*Eugene Weekly*

Riverwalking

Riverwalking

Reflections on Moving Water

KATHLEEN DEAN MOORE

A Harvest Book • Harcourt, Inc.

Orlando   Austin   New York   San Diego   London

www.HarcourtBooks.com

The following essays have been previously published in slightly different forms:
"The Rogue River" in *Willow Spring;* "The Jet Stream" in *Northwest Review;*
"Winter Creek" in *Commonweal;* "The Willamette" in *North American Review;*
"The Metolius" in *Wild Earth;* and "Alamo Canyon Creek" in *Forest of Voices:
Reading and Writing the Environment,* ed. Lex Runciman and Chris Anderson
(Mountain View, CA: Mayfield).

Lyrics from the following songs are reprinted by permission:
"Cheating on the Blues" by Don Cook/Kix Brooks/Chick Rains copyright © 1990
Sony Tree Publishing Co., Inc./Sony Cross Keys Publishing Co., Inc./Fort Kix Music.
All rights administered by Sony Music Publishing, 8 Music Square West, Nashville,
TN 37203. All Rights Reserved. Used by Permission.

"The End of the World" by Arthur Kent and Sylvia Dee copyright © 1962, 1963
(Renewed) by Music Sales Corporation (ASCAP) and Ed Proffitt Music.
International Copyright Secured. All Rights Reserved. Reprinted by Permission.

Library of Congress Cataloging-in-Publication Data
Moore, Kathleen Dean.
    Riverwalking: reflections on moving water/Kathleen Dean Moore.
        p.   cm.—(A Harvest book)
    Originally published: New York: Lyons & Burford, 1995.
    ISBN 978-0-15-600461-9
    1. Nature. 2. Rivers—United States. 3. Human ecology. 4. Moore, Kathleen Dean.
I. Title.
QH 81.M854   1996
508—dc22   96-11364
Text set in Granjon
Designed by Kaelin Chappell
Printed in the United States of America
First Harvest edition 1996
DOH 15 14 13 12 11 10   9 8

To Frank,
and to Erin and Jonathan,
partners on a great expedition.

Contents

Preface

For as many years as I can remember, I have walked in rivers. Each Sunday afternoon, through all the summer and winter Sundays of my childhood, my father led nature walks through the beech-maple forest at the bottom of a valley that divides the suburbs from the western edge of Cleveland, just under the final approach to the Cleveland Hopkins Airport. He was the Rocky River Park naturalist—walking briskly through the woods at the head of a wandering line of citizens, carrying a tame crow on his shoulder, lifting rotten logs to find salamanders, inviting strangers to look through his hand lens at the swollen reproductive organs of woodland flowers. Meanwhile, my mother got the refreshments ready back at the museum, and my sisters and I walked the river.

Rocky River is a shallow, shale-banked stream fed by runoff from the hills south of Cleveland, green-humped hills sheltering shopping malls and dog kennels. Nobody swam in Rocky River because the *E. coli* count was formidable, but there was no harm in putting on a pair of old tennis shoes and walking through the brown water, through floating leaves piebald in broken light below the beeches, along towering shale banks, past teenagers washing their cars in shallow bays and families eating lunch. We always watched for hatchet heads in the water, because we knew that Indian wars were fought for the hills over the river. We broke off pieces of shale and split them in our hands, looking for fossils. There must have been small revelations in the layered surfaces, between sheets of rock, under the sheen of deflected light, but all I remember is the muddy smell of algae and the weight of the water—rich, resistant water pressed into a bow-wave against each shin.

All those years, I desperately envied people who traveled down rivers in boats. Whenever we drove past the wrecked hulk of a rowboat washed up on the Lake Erie shore, I tried to get my father to stop and evaluate the prospect that, with a little paint and a little fixing up, the boat might float. We rented rowboats for birthday treats, because my sisters and I could not imagine a greater excitement. A decade later, my husband and I honeymooned on his parents' island in Canada where boats floated in the pine- and gasoline-scented darkness of the boathouse. And now, all these years later, a continent away in Oregon, our own little garage, built of wood to hold a Model T, is stuffed instead with boats while our car sits out in the rain and grows moss under its bumper. A Grumman canoe, a McKenzie River drift boat on a trailer wedged diagonally across the garage, a small wooden dory suspended from the ceiling and, in a bag in a corner, an inflatable kayak that looks like a banana—in all my childhood daydreams, I never imagined such great good fortune.

All the same, boats are designed to separate a person from a river and now, when I have a choice, I would rather travel down rivers on foot, walking along trails that run the length of the river or, best of all, wading through the river itself. When I walk in a quiet river, I move through a reflection of the landscape. The mudbank, the willow thicket, the mare's-tail clouds lie flat around me, upside down. The river bisects me at the waist— half observing, half immersed in the gently rocking image of the land. When I press forward against the current, the landscape folds and compresses. Next to my body, it breaks into patches of color that ride past me on a spreading wave. To my back, the willows re-form, bend as if in a high wind, then settle and reach out to the reflection of their roots.

Above my head is a world of light and air and swallows. At my waist, the transparent forest. Below that, vaguely visible through a wash of tree and sky, the round stones of the river bed and my own feet in tennis shoes. The laces flow behind, trailing weeds.

The river carries a history of the land and the people who live on the land, stories collected from a thousand feeder streams and recorded in pockets of sand, in the warm and cold currents, the smells of the water, the mayflies. The river carries my own history, swirls of silt lifted by my passage, memories so thick and slippery that I struggle to keep my feet. This is where I walk, sliding on river stones.

By now, my childhood family has scattered and I have a family of my own. I have become a philosopher by training and occupation; my husband is a biologist. We are both professors at Oregon State University, where we occupy offices on opposite sides of a shady brick pathway. From professional lives that fragment our studies into discrete objects and focus our attention on answerable questions only, we come together on rivers where biology and philosophy, body and mind, experience and idea, flow side by side until they cannot be distinguished in a landscape that is whole and beautiful and ambiguous. This book is about what I have seen and heard in that land.

The essays in this collection are river essays because I began to write each one alongside a stream or floating down a river, and so they may still carry the smell of willows and rainbow trout. Drifting on rivers, you know where you will start and you know where you will end up, but on each day's float, the river determines the rate of flow, falling fast through riffles, pooling up behind ledges, and sometimes, in the eddies at the heads of sloughs, curling back upstream in drifts marked by slowly revolving flecks of foam. So, drifting on rivers, I have had time to reflect—to listen and to watch, to speculate, to be grateful, to be astonished.

I have come to believe that all essays walk in rivers. Essays ask the philosophical question that flows through time—How shall I live my life? The answers drift together through countless converging streams, where they move softly below the reflective surface of the natural world and mix in the deep and quiet places of the mind. This is where an essayist must walk, stirring up the mud.

K.D.M.
Corvallis, Oregon

Acknowledgments

With warmest thanks...

To my family, partners on these grand expeditions. To my daughter, Erin, now a senior in college, but once a tiny baby sleeping in a tent on the shore of the North Platte River while snow fell in the night. To my son, Jonathan, a college freshman who appears in my memory as a baby asleep in a tent under beach pines while the beam from the Bandon lighthouse sweeps across him. And most of all to my husband, Frank, a wise and strong man, a maniac fisherman, a man who is fresh air and bedrock to me. How much I love these people, the reader will soon learn, and how much insight I have gained from them, how great a joy I take in their company.

To Chris Anderson, an essayist, and Lex Runciman, a poet, who accepted a colleague as a novitiate and introduced me to the mysteries of the essay; without their encouragement and expertise, there would have been no book. To my current writing group, Gail Wells, Marion McNamara, Steve Radosevich, and Bruce Weber, for all their help, week after week. To Marjorie Sandor and Ted Leeson, for good advice, generously given. To Bob Mason and Dick Jones, who are scientists, for the wild stories about snakes and ducks.

To Todd and Susan Brown, for sharing their rivers, their store of knowledge, and their deep love for wild places. We traveled with Todd and Susan on most of the trips described in this book; riverwalking with them is a comfort and a joy.

To Bill Rieckmann, for generous gifts of many kinds.

To Lilly Golden, editor at Lyons & Burford. I first spoke to Lilly when she was in her New York office and I was standing at a roadside phone in Montana, while water from a lawn sprinkler rhythmically ratchetted around and wet my legs. From that time to this, she has been a source of good will and good ideas.

To the rivers, whose stories I have taken without asking. Wanting to protect your secrets, I have sometimes changed your names. I hope I haven't hurt you.

The life of the mind is not the rotation of a machine through a cycle of fixed phases, but the flow of a torrent through its mountain-bed, scattering itself in spray as it plunges over a precipice and pausing in the deep transparency of a rockpool.

— R. G. COLLINGWOOD

Riverwalking

One

The Willamette

*I wanted my daughter to lie in the tent, pressed between her brother
and her father, breathing the air that flows from the Willamette
River at night, dense with the smell of wet willows and river algae.
I wanted her to inhale the smoke of a driftwood fire in air too thick
to carry any sound but the rushing of the river and the croak of a
heron, startled to find itself so far from home. I wanted the chemical smell of the tent to mix with the breath of warm wet wool and
flood through her mind, until the river ran in her veins and she
could not help but come home again. That is why, on the weekend
before my daughter left for Greece, I made sure that the family
went river-camping on the Willamette.*

My daughter comes from a long line of people with strong homing instincts. My daughter's grandmother, my mother, was born
in a river town, Thornaby-on-Tees, Yorkshire, in a brick house
three blocks from the North Sea. Although she breathed the sea-
weed wind, she never walked on the beach below the limestone

cliff at the end of her street. Warning signs and long coils of barbed wire protected the homeland against German invasion and kept my mother away from the sea.

Her father was one of the few men left at home in England during World War One. He was a ships' model builder who carved little wooden warships, patterns for the shipyards at Sunderland. Sometimes he brought his work home. My mother's strongest memory of England was the fragrance of fresh cedar curling under his plane and falling in long coils on the kitchen table. As he worked the wood, my grandfather sang Scottish folk songs—*You take the high road and I'll take the low road*—and the Navy hymn, *for those who perish on the sea.*

One night, my grandfather took my mother into the city to see where the zeppelins had firebombed the streetcar barns. Under cover of darkness, the zeppelins had moved slowly, silently, upriver to the shipyards, following the lights of the ships on the Tees. They dropped their bombs, turned, and ran back to the sea. After that, blackout was imposed, and the zeppelins, when they returned, followed moonlight reflected on the river. My mother thought how beautiful the river must have appeared to the German airmen, its surface ablaze in reflected light from exploding buildings, and how anxious they must have been to get back home.

After the war, the shipyards shut down and my grandfather—unemployed and uneasy—began to think about emigrating. He argued for America because it was the only country that printed "In God We Trust" on its money. My grandmother refused to leave home unless she could come back to Thornaby-on-Tees every four years for the rest of her life. When my grandfather made her that promise, she gave away everything that wouldn't fit in three tea chests and dressed her children in their best clothes for the trip to America. Their destination was Cleveland, where an uncle had an extra room.

From the day they arrived, the family saved everything they had for four years and then, every four years, spent everything they had saved to go back home. They traveled on the Cunard Line, second class. I have the picture postcards they sent back to America: yellowing photographs of oversized ships lined with tiny people wearing hats and waving. I also have ten teacups, one for every trip. They are decorated with painted ivy or pictures of the Queen. One says, *there'll always be an England.*

When I was first married, we lived in an apartment above a delicatessen in Cleveland. The apartment had a balcony overlooking a major arterial. Evenings, we sat on the balcony in lawn chairs and watched the city life flow out of Cleveland in heavy white cars. One night we drove into the city to see the oil slicks burning on the Cuyahoga River. We couldn't get close enough through the brickyards and factory gates to see the river. But we could see thick black clouds trapped in the rivercourse, glowing red underneath.

We began to think about leaving Cleveland. My husband argued for Oregon, because it had clean, cold rivers. So we moved to the Willamette Valley and made our home here. But I returned to Cleveland every year at Christmastime.

I have a picture in my memory of the drive through darkness to the Portland airport, with white fog flowing down the Willamette River as if even the air ran to the sea. I have a picture of the airport in early morning darkness. People are crowded together with their coats on, some sleeping curled up in chairs, their belongings in piles beside them. Everyone is solicitous, subdued, uneasy, as I imagine people to have been in the harshly-lit tunnels under London during the War.

Fewer and fewer things drew me back to Cleveland each year, but still I went. At first there was the house I grew up in, and my mother and father and my two sisters, and the carols we always sang after supper. We sang slowly, in four-part harmony—deep, rich, thick Methodist chords. My mother chose the songs and gave us the pitch. *Winds through the Olive Trees, four verses.* And there were the English family foods—roast beef and Yorkshire pudding and decorated cookies between layers of wax paper in the roasting pan. We always took a picnic to the park in the snow, turning the picnic table on its side to shelter us from the wind coming off the lake. And always, there was the same joke about how the house was so full of people that we would have to cut off the back half of the Christmas tree and wedge it up against the wall.

But the house dropped out of the picture when my mother and father moved to a smaller place. My sisters got married and moved to bright new houses in the suburbs near Pittsburgh and Baltimore. Then my mother died. We couldn't pitch the songs. My father got sick. And so the rituals dropped away one by one, until there was nothing left at home except my father and the joke about half a Christmas tree. Finally, it seemed that the only reason to go home for Christmas was that someone needed to be there to hear my father tell the joke.

I know a biologist who studies the homing instinct in garter snakes. He says that garter snakes spend the winters clustered together in rock piles underground, ancestral wintering dens that may have been home to the snakes for a thousand years. In the spring, all the snakes crawl out and travel, maybe half a mile. Each one establishes a home base for the summer, a pile of leaves or the space under a fallen log. They travel out each day, but each night they make their way back home along the same trails. When biologists draw a snake's wanderings on a map of the land, the lines are thick, drawn back and forth, back and forth, like rays extending out from a home base. In the fall, the adults travel back to spend the winter with their relatives in their ancient family home.

Before they return to the den, the females give birth to a pile of lithe little snakes and then move on, leaving the babies to fend for themselves. The babies spend the winter, who knows where. But the next fall, the yearlings travel back unerringly to their ancestral home—a place they have never been. As they go home, they will pass over other dens that would be perfectly good places to spend the winter, not stopping until they get to the den that shelters their own elderly aunts and distant relations.

Scientists know so much about homing in animals: Bees orient to polarized light. Salamanders steer by lines of geomagnetic force. Garter snakes follow scent. Pigeons use the position of the sun. Songbirds follow the stars. They are all drawn to a place proved to be safe by the hard, undeniable fact of their own existence.

But who has studied the essential issue: What will draw our own children back home?

By the time we got all our camping gear stowed away in the drift-boat, it was late afternoon. In the shadows of the riverside cotton-woods, the air was cold and sharp. So we drifted along the eastern

bank of the river, glad for the warmth of the low light. We pulled
up onto the gravel beach of an island thick with willows and set up
the tent on a pocket of sand.

After supper, my daughter and I walked down the shore. We
wore high black rubber boots and walked sometimes in, sometimes
out of the water, the round rocks grinding and rolling under our
feet. Far ahead, a beaver slapped its tail against the river. We talked
quietly—about her visa, about loneliness, about how the skyline of
the distant coast range seemed to glow in the dark.

Fog thickened the darkness, so even though it wasn't late, we
turned back toward our supper fire. We didn't talk much on the
way back, but we sang like we often do along the edge of a river,
where the density of the air and the rush of the river make the music
rich and satisfying. We sang the Irish Blessing—*my daughter sang*
the soprano part—and we did fine, the river singing the bass line,
the rocks crunching under our boots, until we got to the last bless-
ing: May the rain fall softly on your fields. Then I couldn't do it any
more. I sent my daughter back to the fire alone. I lay face down on
the round rocks and cried until the steam from my lungs steeped
down into the dried mud and algae, and the hot breath of the river
rose steaming and sweet around my face.

Maybe the homing instinct is driven by traditions: hanging
Christmas stockings each year on nails pushed into the same

little holes in the mantel. Maybe it is driven by smells or tastes or sounds. But maybe the homing instinct is driven only by fear. On the road, at dusk and away from home, the foreboding, the oppression of undefined space, can be unbearable. Pioneers knew this dread; they called it *Seeing the elephant.* Starting out, the wide open spaces were glorious—the opportunities, the promise, the prairie, all fused with light streaming down from towering clouds. Then suddenly the clouds became an elephant, a mastodon, and the openness turned ominous. The silence trumpeted and the clouds stampeded. Dread blackened the edges of the pioneers' vision. They saw the elephant and turned their wagons around, hurrying through the dusty ruts back to St. Louis. They had to go back. They had to get home.

The French existentialists knew that feeling: *la nausée,* existential dread. The pioneers—they, we—walk out into a world we think makes sense. We think we understand what things are and how they are related. We feel at home in the world. Suddenly, without warning, the meaning breaks off the surface, and the truth about the world is revealed: Nothing is essentially anything. The prairie gapes open—"flabby, disorganized mass without meaning," Sartre said. Pioneers can create meaning by their decisions, but those decisions will be baseless, arbitrary, floating.

This discovery comes with a lurch, thick in your stomach, like the feeling you get when you miss a step on the stairs. When the feeling comes over you, you have to go home, knowing that home doesn't exist—not really, except as you have given meaning to a place by your own decisions and memories.

Robins singing woke me up in the morning, a whole flock of robins at the edge of the Willamette. Each robin was turned full into the sun. I climbed out of the tent and sat cross-legged on the gravel, my face turned toward the warmth, my eyes closed, bathed in pink light. Soon my daughter, in long underwear and rubber boots, ducked out of the tent and walked into the river to wash her face. She scooped up a pot of river water and carried it to the kitchen log to boil for tea. Crossing to the campbox, she rummaged around inside until she found matches, scratched a match against a stone, lit the stove, and set the teapot on the burner. Then she sat on the broad log in a wash of sunlight, pulling her knees up to her chest and tilting her face toward the light. Her hair, in the sun, was as yellow as last winter's ash leaves in windrows on the beach.

Scientists say that a wasp can leave its hole in the ground, fly from fruit to fruit, zigging and zagging half the day, and then fly straight home. A biologist once moved the three rocks that framed a wasp's hole and arranged them in the exact same pattern, but in a different place. The wasp landed between the rocks, right where its hole should have been, and wandered around, stupefied.

My three rocks are the Willamette River. Whenever I walked out of the airport, coming home from a visit to my father's house, I could smell the river, sprayed through sprinklers watering the lawn by the parking lot. The willow-touched water would wash away the fumes of stale coffee and jet fuel and flood me with relief. This is what I want for my daughter.

The John Day River

We're driving fast through Oregon's high desert country, listening to country-western music turned up loud. This is dancing music, Texas Two-Step, gleeful songs about broken hearts. *Have some fun, keep it light/Cheatin' on the blues tonight.* Frank taps the steering wheel in time to the music. The boat swings along behind us. It's Frank, the kids, the boat, and I, two-stepping across the plain, swinging out of Prineville, *rollin' on down the road.*

Warm air pours through the van, carrying the smells of sage, juniper, roadkill skunk, asphalt, juniper again, sage. We're moving through land made of huge basalt slabs layered like buttered pancakes in a sliding pile. Junipers grow in the shadow of the rimrock. From the base of the cliffs, the hills curve down into distant valleys where cottonwoods line the creekbeds.

We swing into a gas station for a tank of gas and a block of ice. I say "howdy" to the gas station attendant and he howdies me back and laughs. Frank walks round behind the van and snaps the tie-downs on the boat.

We are headed for the John Day River, for a seventy-mile float trip between Harney and the Three Springs Bridge. We should get to the put-in at high noon, and this afternoon we will run Harney Rapids. Tonight, *good Lord willin' and the creek don't rise,* we will slide our sunburned bodies into cold sleeping bags, tired out from flinching from water that splashed in our faces, water that smashed back from rock cliffs at the turn of the river, tired out from riding in the boat—bucking, rocking, and finally sliding out the bottom of the rapid, our clothes heavy with cold water, the tops of our heads hot from the sun. Tomorrow the river will flatten out, and we will float for three more days, riding through a deep canyon past sand beaches rimmed with willows, past rough cliff faces pocked with the clay-jar nests of swallows.

The pleasure of the plan makes me think of John Stuart Mill, who was a philosopher specializing in happiness. He said that *the main constituents of happiness appear to be two. Tranquillity and excitement. With much tranquillity, many find that they can be content with very little pleasure. With much excitement, many can reconcile themselves to a considerable quantity of pain.* If so, a good river must be essential happiness, happiness distilled and running between high banks, because on a river, tranquillity and excitement alternate every half mile, every tight, cliff-bound curve, every quiet pool where white flowers float on the reflection of the sky. I think it's a pity that Mill, a nineteenth century philosopher in London, never found a way to run a desert river.

The John Day River was named after a Kentucky hunter who came into the Oregon country with John Jacob Astor's men in 1810. He was destined to measure the precise quantity of pain to which a man can reconcile himself. Day was a strong-backed, leather-hearted man, but he had lived his life *too fast*—his words—and at the Idaho border, he wore out. The other men in the company, already out of food, facing winter, left Day and his partner, Ramsey Crooks, on the shore of the Snake River. As they tried again and again to push through the Wallowa Mountains in the winter, Day and Crooks sometimes found horseflesh to eat and old beaver pelts they could chew. But mostly they had nothing to eat at all. They collected an old fish skeleton and ate its bones, pounded to powder. They gathered roots, thinking to boil them, but by then they had lost their flint. When they finally made it to the Columbia River, a band of native people captured them, beat them, and stole their clothes. Astor's men found Day and Crooks naked at the mouth of the river, starved, helpless and hopeless, screaming to be saved. When he came back up the Columbia with a trapping party in the fall, John Day lost his mind. The men ferried him back to Fort Vancouver, tied in the bottom of a boat. He died within the year. From that time, John Day's name has been linked to the river where he was rescued.

We pull the boat north into the Ochoco Mountains, where yellow ponderosas shade the road, and the breeze smells like springtime snow on mountaintops. When the road turns east, sunshine blinks through the pines, flashing in 4/4 time.

At Mitchell, we stop beside a pay phone under a row of poplar trees alongside the road. I need to call my dad, sick in a hospital in Ohio, to check in, to make sure he will be all right while we are incommunicado in the river canyon. Midway through the medical reports, a golden eagle floats into the top of a poplar. It chitters and beats its heavy wings as it settles into the branches, keeping one eye on the road. Huge, powerful, unexpected, the eagle makes it hard for me to concentrate on my dad. I speak consolingly even as I gesture wildly, trying to get Frank to look.

Back on the road, Frank accelerates hard up a hill and over the top. *Oh I can't understand/No I can't understand/Why life goes on the way it does.* It's a cowboy waltz. One two three. One two three. Outside swing. Inside swing. Frank pulls the boat around the broad curves. He's a good driver; he keeps things under control while they're spinning.

We leave the basalt slabs and enter the Painted Hills, a land of bare little hillocks striped in lemon yellow, pink, orange, lime green, chocolate—successive layers of ash blown out of distant Oligocene volcanoes. Closer to the John Day River, the exposed hillsides are sea blue clay, dripping down the canyon walls like melted wax. Fossils are buried in that soft, slippery blue ash: giant rhinoceri, saber-toothed tigers, turtles as big as saddles, camels, piggish oreodonts, gigantic go-for-the-gusto animals that gambled everything on their huge size and lost everything in a prehistoric volcanic explosion.

At the put-in at Harney, we bump down a dirt road onto the gravel and walk to the edge of the John Day. The river is broad and brown, in flood, 4,700 cubic feet per second at the gauging station by Service Creek. Is that too much water? Is it a go? Two miles downriver at Harney Rapids, tons of water will be driven between two hills. There the river will slow, hunch its back, swell up and finally pour with terrific force in one wave over the falls. It takes a strong person to hold the oars against that much hydraulic pressure. We decide to make the run, but at Harney Rapids, women and children will walk.

I'm one of the votes for caution, because I go through life slightly unnerved by Ralph Waldo Emerson. I half believe him when he says that *Every sweet has its sour; every evil its good. For everything you gain, you lose something. Nature hates a monopoly.* This makes sense to me. I think it's at least plausible that everyone's life finds its own equilibrium, a natural balance of joy and pain. But if happiness has to come out even with sorrow in the end, then I am in big trouble. I try to take my joy in tiny sips, hoping my sorrow will be equally shallow. But usually I end up swallowing happiness in gluttonous gulps, believing all the while in justice, and so rejoicing at little setbacks and petty unpleasantness, courting small disasters, hoping to eat away at the deficit. I know that if the universe is as reciprocal as Emerson says, I have already had more than my fair share of happiness, and in the end I will have a tremendous, bone-crushing debt to pay.

Two miles downriver from the launch, Frank slides the driftboat onto the beach above Harney Rapids. The kids and I climb out and pick our way over the bluff to a vantage point above the rapids. From a rock high above the river, we watch Frank tie down the gear in the boat, rearrange it, and tie it down again. Swallows swarm like horseflies around the face of the cliff. We see a canyon wren that is surely singing its heart out, but we can't hear anything above the booming of the river. From this viewpoint, we overlook a full mile of river—the frothy, broken upper rapids, the smooth heavy slide down Harney, and then the rocky chop below the falls.

Frank pushes off, the current catches the boat, and he begins to pick his way through the rocks with long strokes on the oars. I walk away from the edge of the bluff. I really don't want to watch this, but I sit where I have a clear view of Jonathan, who has not taken his eyes off Frank. For the longest time, Jonathan does not move. Then his face compresses and crumples, and he is running down the hill to the base of the falls. Erin and I catch up to him at the shore of the river.

Frank's boat is tilted on edge under the falls, jammed sideways against a wall of brown water. The force of the water presses down on an oar and submerges the starboard gunnel. Standing up in the boat, Frank heaves the oar out of the oarlock. The boat pops away from the falls and spins down the river, thrown from rock to wave, while Frank struggles to release his spare oar and we run as fast as we can down the beach.

In a quiet eddy at the bottom of the rapid, the long oar floats round and round. While we wait for Frank to ride down to us, we can hear the canyon wren's song against the hiss of the river in the sedges, a soft descending scale. Awkwardly, with uneven oars, Frank pushes toward shore and we wade out to haul him in. Jonathan clambers over the side of the boat and sits tight against Frank, quiet, close, while I pull the anchor up onto the sand.

Frank ships his oars and loosens the straps on the ice chest. Soon we are sitting side by side on warm sand, drinking beer, eating lunch, laughing. The kids tell the story of Frank's adventure over and over, and each time they think it's funnier.

I am uneasy, awash with relief, soaked in sunshine and pleasure. Where does this happiness come from? What is its cost? I think about how close Frank came to paying off part of our debt in Harney Falls. I think about my phone call to my dad, and it suddenly strikes me that there are other ways that a vengeful Nature could even out the sum total of pleasure and pain in the world. Maybe *one* person's happiness is balanced with *another* person's pain. When cowboys pull their pickup trucks into an intersection in the middle of the wheat fields and turn up their radios to dance in the headlights, maybe that forces Nature to crank up the pain of another person who lies alone in a hospital bed. Or maybe I have the cause and effect reversed: Maybe suffering in one time causes joy in another. Our celebration on the beach may complete a process of justice that began when John Day cowered naked between the rocks where the river empties into the Columbia. Or maybe our joy is the cosmic complement of a doomed rhinoceros stampeding away from a suffocating sea blue cloud of ash. Or is my father's pain paying for this trip?

We load up the boat and push into the current. Sitting on the bow, our feet dangling over the water, we float between canyon walls and sing. *I got the blues tonight.* Ahead of us are three sun-drenched riverdays and a calm, curving river.

The Rogue River

> "My proposal is that we should deliver a full-dress
> oration in praise of Love, each one from left to right, and
> Phaidros must begin first, since he is in the first place."
> Then Socrates said, "No one will vote against you,
> Eryximachos. I could not refuse myself, I suppose,
> when love is the only thing I profess to know about."
>
> —PLATO, *Symposium*

ROGUE RIVER VALLEY, OREGON,
9 PM, JANUARY I

Because the moon was full and closer to the Earth than it would
be again for seven years, Frank and I went skiing at night along
the headwaters of the Rogue River. We headed west, the moon
at our backs. Our shadows dug dark holes in the snow, forcing
us to ski on the edge of crevasses that glided away in front of us.
Alongside the trail, the snowbanks glowed from underneath

and lit the bottom branches of the trees. When we turned north, the fir trees cut the moonlight into wide ribbons, so we skied across alternating strips of snow and dark space—slower in white air, faster in dark air, gliding in moonshine, airborne in shadows, until we slid down the last slope and came to rest on a bridge over the Rogue.

That high up in the mountains, the river was narrow and black as outer space. It fell in slow curves from the moon, under the bridge, and over the edge of the earth. Needles of fog frost covered everything touched by air rising from the water—the railing on the bridge, the cattails, the snow mounded along the banks. With our ski poles, we pushed crystals off the edge of the bridge. They fell like a meteor shower and winked out in the water.

In ancient Greek stories, Eros was a dangerous God: Eros, the God of Love, the enemy of Reason, treacherous, powerful, impossible to resist. Under the control of Eros, Helen eloped to Troy and Medea betrayed her father. Cassiopeia was tied in a basket and hung in the sky, a vain queen driven by love to evil and then ruin. Passion, passive, *pathos,* pushed, over the edge.

We say that people build relationships. They work on their marriages. They grow together. But they fall in love. *Je suis tombée amoureuse.* In Chinese, "I fell on love." In motion. Out of control. Against all reason.

BERTHOUD PASS, COLORADO,
10 PM, FEBRUARY 21

"It is ten o'clock at night," Frank said. "We have skied. With full packs. For seven miles up a mountain trail. We are already in our sleeping bags. In a hut. With a woodstove. The snow outside is twenty feet deep. The wind is blowing. And now you've decided you're not spending the night."

"Right."

"Why the hell not?"

"Because it smells mousy in here, and because we're missing out on the stars."

Frank went to the door, opened it, and for a long time looked out into the night. Then he walked back toward the woodstove, where his clothes were warming by the fire, and started to pull on layers of long underwear and socks. I put on my boots and everything else in my pack that was wool or goosedown, and we went outside. There were stars in every direction—galaxies thick in the sky around us, glittering in the snowbanks. A small wind lifted stars off the snow, sending them spinning over the glazed hilltop. We shook out the tarp and laid it flat on a rise 100 feet from the cabin, above treeline, with a 360-degree view off into space. We laid our foam pads and sleeping bags on the tarp and slid in, fully dressed. Frank took the windward side and tucked me into his lee. When we had pulled the edge of the tarp up to our chins and drawn the sleeping bags tight around our faces, we lay on our backs and stared straight into Orion's belt buckle.

At the end of the night, a full ring of lavender light flickered around the edge of the world. Lavender streaks flared up into pink, then peach, then yellow, spreading, until we were surrounded by a ring of fire. Willow ptarmigan started to yelp from every direction. The more the morning flared up, the more frantically they yelled. When the flames met in the middle and the whole sky blazed, the ptarmigan calmed down and went about their business. We got chilled, lying on the snow, in the morning. So we dragged all our gear into the hut and went to sleep.

Frank says that the Greeks had it wrong. Love isn't dangerous; love has survival value. If love has to do with forces at all, the forces are softer, instrumental, the gentle sloshing of acids in the stomachs of ancient predators, dissolving the genes of animals who failed to love, when love would have saved their lives or the lives of their children.

The love of a mother for her infant is inevitable, unsurprising, Frank says. Take goats, for example. Whatever a mother goat smells right after she gives birth, that becomes the object of its goofy mother love. This is why, if a tiny kid dies, a goatherd will skin it and tie its steaming hide onto an orphan goat. The mother doesn't know or care. She doesn't love her kid. What she is attached to, what she is compelled to nourish and protect, is the blood-touched smell of a newborn kid: chemicals washing silently over receptors at the nose-edge of her brain. Love is a matter of hormones and pheromones and reproductive necessity, of rhythms and cycles, life and death, chemicals ebbing and flooding like tiny tides under a microscopic moon.

But what about lovers? I ask. What about tundra swans who mate for life and languish and die if their mate dies? What about ptarmigans, who follow each other around, clucking softly, fussing over each other, sifting their feathers? It's evolution again, according to Frank. Swans and ptarmigans nest on the ground, where their chicks need the protection of two parents, one to sit on the nest and one to warn off predators. Wandering foxes have long ago eaten the offspring of faithless parents.

BAY OF ISLANDS, ONTARIO, CANADA
2 AM, AUGUST 15

It was the middle of the night when we finally drove up to the dock at Little Current. We sat in the silent car and weighed our options. We could sleep in the parking lot and wait for the boat to come for us in the morning, or we could untie the canoe from the top of the car and paddle across the lake to the island that night. The wind was calm, the air was warm. We had made the run many times before, in the daylight. We had been in the car all day and all night. So we slid the canoe into the water and headed in the general direction of Cassiopeia.

All we could hear were water sounds, and all we could see were stars—stars in the black sky, stars on the shining black surface of the lake. We couldn't see the islands, but we could see where stars were not reflected on the water, so Frank steered to avoid the empty spaces.

Before long, the stars started falling. For each star that arched toward Earth, another star flashed up from the bottom of the lake. They collided and annihilated each other without a sound. They fell from all directions. Some fell straight down and vanished. Most came coursing in from Perseus and fell rocking in our bow-wave. We checked the sky to see if any stars could be left. We listened for trumpets. All this glorious disaster and not a sound but the canoe pushing softly through warm water and Frank, quietly reminding me to paddle.

Socrates thought that love starts when people are attracted to the beauty of others—the form of a face, a pair of hands, a voice. As the years go by, they discover an even greater beauty in the people they love, a loveliness that cannot be seen, the beauty of the soul. Drawn toward this invisible beauty, they come to love the idea of beauty. And so gradually, "ever mounting the heavenly ladder, stepping from rung to rung," the attraction of beauty leads people to a vision of absolute Beauty itself.

I get nervous when philosophers start to capitalize their nouns and talk about absolutes, and Frank has even less use for this kind of talk. Still, I think Socrates had it mostly right: Beauty can lead people to love. And I think it works the other way, too: Love can lead people to beauty. At least this has been our experience.

MAGRUDER BAY, ONTARIO
11 PM, AUGUST 28

Sturgeon fishing is best after dark, so it was deep night before we quit fishing, started up the outboard, and headed back to camp, and by then we had a bit of a problem. We weren't lost, strictly speaking, because we knew where we were and we knew where camp was, but a string of small islands lay between here and there. On the way to the sturgeon hole, we had found a channel that was deep enough to allow passage through the islands, but now we weren't sure exactly where that channel was, and the night was very dark.

Mulling over our prospects, we surveyed what we had in the boat: fishing tackle, a flashlight, a can of beer, and a pail of worms. It would be a long night if we didn't find camp. Nothing to do but try one passage after another, until we found one that would let us through.

We motored slowly down the string of islands, passing through pockets of warm air that smelled of white pines. Frank made a wide turn and motored back up again, slower this time. He trimmed the engine almost to stalling and turned toward the islands. The *putt putt* of the engine bounced off a rock wall somewhere to starboard.

I leaned over the bow with the flashlight pointed down into the water and watched for rocks. From the void that surrounded us, the flashlight's beam created a tube of daylight sea as clear and as green as bottle glass. I could see the lake bottom passing slowly through a circle of light—a boulder slicked with algae, a bed of pebbles in all the colors of the rainbow, narrow ribbons of grass waving slowly in green light, a mussel shell. Suddenly a bank of rock porpoised out of the lake.

"Cut it."

Frank killed the engine. Quiet flooded into the space where the engine's noise had been, and again the smell of pines came across the water. After a few moments, our wake rustled against the gravel beach, and then we drifted in silence, wondering what to do next. Suddenly I realized that where I had seen only darkness, I now could see bright water edging onto the pebbles in a rocky cove.

I looked up. The sky above the northern horizon was flushed with green. Beams of light roved through the brightness, fluttering high above the horizon and ruffling the sky. Liquid flares of green light spurted into the darkness and flamed out, leaving an oily sheen across the sky. The sheen condensed into a cloud and poured toward earth, a sheeting rain of green light that seemed never to stop or to be replenished. Pink mist ascended through the great chiming light. Frank stood up and yanked on the starter cord. The universe lurched violently and filled with sound.

Slowly, slowly, we skirted a rock outcrop and, by the light of the aurora, picked our way through a channel so close to an island that I could run my fingertips along its granite ledge. Then we were on the other side of the islands with the glossy water of the bay stretching out in front of us, the radiance overhead and, two miles across the bay, a single white light on the dock at camp.

Winter Creek

I wish to speak a word for the art of poking around. Although the art can be practiced in libraries and antique stores and peoples' psyches, the kind of poking around I am interested in advocating must be done outdoors. It is a matter of going into the land to pay close attention, to pry at things with the toe of a boot, to turn over rocks at the edge of a stream and lift boards to look for snakes or the nests of silky deer mice, to kneel close to search out the tiny bones mixed with fur in an animal's scat, to poke a cattail down a gopher hole.

People who poke around have seeds in their socks and rocks in their pockets. They measure things with the span of their hands. They look into the sun when they see a shadow pass across a field. They spit in rivers to make fish rise. When no one is looking, they may even rub their lips where beavers have chewed, just to get a sense of it. Often they stand still for a long time, listening, and then they follow the sound, sneaky as a heron, until they are close enough to see a chickadee knocking on wood like a tiny woodpecker. But if the route to the chickadee is crossed by the tracks of a black-tailed deer, they will turn to follow the deer into the firs, unless the deer tracks cross a creek, in which case it is important to meander with the water through the fold between the hills.

Poking around is more capricious than studying, but more intense than strolling. It's less systematic than watching, but more closely focused. Unlike hiking, it has no destination. Above all, poking around is not as serious as walking, the noble art so eloquently advocated by Henry David Thoreau. This is because poking around doesn't take much sustained thought, whereas Thoreau insisted that you must think while you walk, "like a camel, which is said to be the only beast which ruminates while walking."

When you walk into a field of reedy horsetails, you could deduce that there is groundwater under your feet and remark on the fact that you are in the presence of a plant as old as the dinosaurs. If you wanted to, you could imagine huge lizards chewing the stalks and blatting like french horns to claim ownership of the marsh, and you could formulate a theory of mass extinctions and apply it to humankind. But if you're only poking around, you might prefer just to cut a section of the hollow stem and blow across it, trying to make it hoot like the mouth of a beer bottle. You could study the flatness of a skipping stone and imagine layers of mud drying in the sun, but if you're just poking around, you'll probably spin the stone into the air instead, trying to make it drop into the stream at such a perfect, vertical angle that it disappears with a small sound like wood on wood, raising a blister on the water.

Of course, there are no rules about this, and some people prefer to keep their minds engaged while they're poking around. If so, the most fitting kinds of mental activity I have found are wondering and hoping. Knocking snow into the creek, I wonder why it turns clear before it melts. I wonder where the gopher went after he raised up a lumpy mound of earth in the pasture last night. Stopping to eat my sandwich in the depression where a deer has slept, I really hope there aren't ticks. I sit on a damp log in the ash swale and wish that the varied thrush would whistle again before my pants soak through. Most of all, I hope that the late winter sun will drop below the oaks and warm my back.

The people who are best at poking around are by no means the people who work the hardest at it, and what they find is often what they least expect. Children are often good at poking around, and among children, Erin and Jonathan are premier, if I may say so. One day Erin lay on her back in the weeds by Winter Creek, blowing across a blade of grass held tight as a reed between her thumbs—idly squealing the day away—when a red-tailed hawk banked into a tight curve overhead and dropped low in the sky, screaming a territorial call that sounded just like air across a blade of grass. Another time, Jonathan played with a pair of pliers in the dark, trying to work loose the wire that held the latch shut on a lantern, not even noticing that the pliers squeaked each time they opened or closed. When he finally looked up, he found himself looking into the face of a tiny screech owl perched in a pine above his head, its eyes intent on the pliers.

When I was growing up, an entire day of poking around was a treat reserved especially for birthdays. My birthday came in the Ohio summer, and my treat was to have the whole family pile into a rented rowboat and poke around Hinckley Lake all afternoon. I can picture us still: Because it is my birthday, I am rowing, dragging strings of elodea off both oars. My father leans over the bow, looking for snails and tiny floating ferns among the duckweed. In the stern sits my mother, rejoicing beyond reason at watery smells; beside her, my little sister on her back, watching the buzzards, and my older sister with a pocketknife, sawing and sawing at the stem of a water lily. The day is aimless, usually complicated by afternoon winds, always unproductive. But each small, individually wrapped observation is a gift.

All this may make it seem that poking around is the perfect avocation for the dissolute, but, on the contrary, it's an art that takes a certain stubborn resolve. You have to drive out ideas that will dampen your spirits or dim your vision: a desk on Monday morning, the dentist's bill. You have to be alert or you will find yourself sucked away by a work ethic as strong as a vacuum: You will stop to pull a single blackberry vine, which will make you look for another and another, until you are dogging from one vine to the next and then heading back to the shed for a shovel. You have to have a strong character, or guilt will overcome you when you realize you forgot to thaw dinner, and you will get back in your car and go to the grocery store and that will be that.

Like most pleasurable activities, poking around has its solemn enemies. Thus parents call their children *slowpokes* and tell them to quit poking around or they will be left behind. Grown-ups who poke around are dismissed as *childlike,* although how that can be an insult is beyond my comprehension. Thoreau says that in the Middle Ages a saunterer was considered a *saint de terre* as he wandered the countryside gathering alms to fund a crusade to the Holy Land, but he didn't go on to say that if someone was exploring the countryside just for the joy of the exploration, he was insulted as a *poke-easy,* a lazy one. Feudal dog handlers took pains to train their dogs not to poke around. If a dog left the scent trail, even to follow the red herrings that the trainers dragged in sacks across the road, the dog was beaten with sticks. The same sort of thing sometimes happens these days to science students, who aren't allowed out in the field until they have reduced the scope of their questions, whittled away at them until the questions are narrow and pointed.

Yet poking around is a guaranteed way to learn. Ideas, after all, start with sense impressions; and all learning comes from making connections among observations and ideas. Insight is born of analogy. Everything interesting is complicated. Since truth is in the details, seekers of the truth should look for it there.

Besides, poking around is recreation, re-creation, in the most literal sense. John Locke said that what gives each person his or her personal identity is that person's private store of recollections. If so, then people should be careful curators of the assortment of memories that they collect over the years. Every time you notice something, every time something strikes you as important enough to store away in your mind, you create another piece of who you are. If someone asks, "Who are you?" it is not enough to say I am Kathy, or I am a professor, or I am Dora's daughter or Frank's wife. The complete answer will acknowledge that a person is partly her memories: I am a person who remembers a flock of white pelicans over Thompson Reservoir, pelicans banking in unison into the sunlight, banking into the shadow, flashing on and off like a scoreboard.

But I don't want to make too much of the instrumental value of poking around. The whole point is that poking around is good in itself, like music, or moonrise. So I poke around at the frozen edges of Winter Creek in the late afternoon when the sun comes in low over the oak knoll and throws a long, rippling shadow from each dried cattail across the creek and up the farther bank. As Thoreau observed in *Walking,* the sun shall "perchance shine into our minds and hearts, and light up our whole lives with a great awakening light, as warm and serene and golden as on a bankside in autumn."

Two

Two

The Little Stoney River

The trick to crossing a river when the current is strong is to take only one step at a time and to make sure that one foot is firmly planted before the other leaves the ground. I hook arms with Frank, get a good grip on a wading stick, and step off the bank. I lift one boot and move it forward. It slides off a rock and wedges. I lift the other boot. The current takes hold of my leg and swings it downstream. I pull it back slowly and push it forward until it slides into a depression between two rocks and holds. I lift the other boot. This time I brace it against the current and slide it along the bottom. How much farther do I have to go? Stones fill the riverbed, stones the size of heads and fists, stones rolling on slabs of granite just broken off the cliffs. Each stone is slicked with algae. If I don't watch my feet, I won't find the likely footholds. But if I look into the river, the sweep of the water carries my eyes with it and almost pulls me on my face. So the only thing to do is feel my way blindly across the bottom, groping with heavy boots and concentrating on the far shore.

I wish I were denser. I wish I were on the other side of the river. "You don't have to cross the river," Frank tells me (dishonestly). "You only have to take one step after another." And so we move diagonally forward, step by step. I take the last three steps in fast, sliding lunges and hit the far side on my hands and knees. I hate crossing rivers, hate it that my legs are buoyant and my balance precarious, hate it when things rush by me this fast and I have to depend for my footing on rocks that have not come to rest. A little mountain creek was what I had envisioned from the line on the map, maybe sparkling and strewn with mossy stepping stones—but certainly not this thunderstorm-swollen, spitting, tarnished silver river.

Frank hitches up his shorts and goes back for the packs and the kids while I dry out in hot grassland on the south bank, wringing out my socks, glad to have that part over with, wondering if the river is rising or falling. We can't stop long, because we don't have much time if we want to make it all the way up the gorge to the head-wall in the high reaches of the mountain. We have wanted to do this hike for a long time, but the mountain is a good day's drive from home and it's hard to find enough days to get there, do the hike, and then get back home in time for work, for school, for soccer, for friends, for piano lessons, for all the rest. This time we have three days, and that is probably enough—three days to climb through five million years of rock.

More than a century ago, Henry David Thoreau said, "Time is but a stream I go a'fishing in," as if he could visit or leave the flow of time at will, like a tourist sitting on the bank with a picnic basket, "wetting a line" now and then, as they say. This is an astounding idea, but it couldn't possibly be true. Time bowls me over, knocks me down, rolls rocks against my ankles. How could time treat Thoreau so gently and be so rough with me?

Eleven million years ago, the place where this great river gorge is now was a flat plain of lava exuded from cracks in the Earth. But for two million years, the entire force of the North American plate pushed against the plain from the east, and the floor of the Pacific Ocean pushed against the plain from the west, so the whole crusty countryside lifted in the center and cracked in long fault lines, the way a slab of gingerbread would rise and crack if you pushed on both sides. The broken edge of one of those cracks lifted a mile high, making the mountain. Then, nine million years ago, when the continent cooled, snow built up slick and heavy on the mountain, compressed into ice, and slowly slid down the mountainside, scraping against bedrock, dragging boulders. In the warmer air at the bottom of the slope, the advancing edge of ice melted and dropped the boulders in a heap, like the pile of coats little children leave inside the front door. In four successive Ice Ages, glaciers formed, scraped down, and melted, gouging a narrow steep-sided canyon a half-mile deep into the sloping side of the fault block, cutting the valley we call the Little Stoney Gorge.

The gorge is splendid, immense. Viewed in cross-section from the plain, it drops in a grand sweep from the snow-bound rim to the sagebrush flat and swoops back up again, a powerful parabola, the trajectory of a nighthawk. The mountain itself is sere, but the cut edges of the gorge glisten, green with aspens and ancient groves of juniper.

Near the bank where I sit, watching Frank and Erin pick their way across the river, a water ouzel works water swirling between stones. The ouzel is a restless, wren-like bird, a bird people call a 'dipper,' because it dips up and down, bending its knees, always in motion, always within the riverbanks of mountain streams, darting from mossbank to splashed rock, dropping into the river to fly submerged along the bottom, bobbing on midstream rocks, genuflecting without rest. Since dippers rarely leave the rivercourse, they live in a world of constant motion, continual uproar. I wonder whether the dippers' regular, controlled movement is their way of maintaining equilibrium, of assuring themselves that it is the river that is rushing downhill, not they who are rushing uphill. Maybe if dippers held still and fixed their eyes on the river, they would tip over on their backs and be swept away in the current.

After the Fourth Great Melting, but before the Fifth Ice Age, we start up the path, carrying our packs. The higher up the gorge we climb, the worse the trail becomes. Again and again, we have to pick our way through boulders and broken rock where a shoulder of the valley wall has slipped off and piled at the bottom of the cliff. Some of the slides are old. There, mountain mahogany and junipers grow up between the rocks and hide the way. But many of the slides are fresh and raw. We trip over mud that has hardened in sharp ridges and we roll our ankles on stones in unruly piles.

Just past one of these slides, we come upon a rock as big as a cabin, sitting smack in the middle of the trail in a muddy ditch that looks as if it had been dynamited out of the ground. I run my eyes up the canyon wall to a raw-edged, pale socket on the rock face. That rock cracked off the cliff, fell half a mile, and bounced one big quarter mile before it plowed through the mud, digging a trench ten feet deep, throwing mud onto trees a hundred feet away. It must have made a fearsome, echoing thud, as the sound waves measured the distance across the canyon again and again and again.

The rock was proof, if any proof was needed, that solidity is only a function of time. A river revealed in a flash of lightning is as thick and quivering as gelatin. And yet, measured against a millennium, a mountain melts down the sides of the valley and pours into the sea.

Time jerks and stalls up here on the mountain. The children hike out of sight, water carves away the edges of a slab of granite, the mountain rises and expands, a mayfly lives out its twenty-four hour adult life, my watch says (nonsensically) 4:30, and I stop to fix my boots because I can feel a blister forming on my heel. Einstein was right: Time can't be a constant. Its speed depends on where you are and the people you are with. This throws me a little, like trying to walk on a conveyor belt that has slipping gears, or stepping off the moving sidewalk at the airport, loaded down with luggage, or trying to cross a river after a storm.

A hundred years ago, a creationist tried to prove that the Earth could not be four billion years old. If the Earth were that old, he said, erosion would have had time to smooth out all the mountains and silt in all the valleys. The Earth would be as smooth and glossy as a billiard ball, which it clearly is not. *Quod erat demonstrandum,* he said. Except for one thing: Even as mountains erode, they are heaved up by forces under the crust of the Earth. The Grand Canyon wasn't created by a river cutting through a plain, but by a plain rising under a river. The important fact about Sisyphus wasn't that his rock kept rolling down the hill, but that he kept pushing it back up again. If the creationist had known uplift as well as he knew erosion, he could have come to know a Creator who is still at work, instead of a sculptor who cast his work in clay and walked away.

Erode. From the Latin *rodere*, to gnaw or chew; from the same roots as "rodent," an animal that gnaws. Erosion is measured in units of time: cubic feet per year. And time is measured in units of erosion: sand running steadily through the narrows of an hour glass, a woman growing old.

A mountain is movement over time. Seismographs draw the movement of mountains, thick vibrations on lined paper, like chords on a bass clef, but slow, *largo*. Hard massed lines, then a thin line sketched over time, then *pizzicato* as a boulder bounces down a cliff. The seismographs look to me like the drawings of birdsongs in the margins of field guides, where time is measured along the horizontal axis, the frequency of vibrations is measured along the vertical axis, and the song itself is a curving line or an ascending scribble. On a sonogram, the liquid whistle-and-warble of the dipper—*tswee, tswee, tsooptsoodle-oop*—is two curved lines and a lumpy smudge. If the dipper's song were inscribed on the horizon, it would trace a lyrical ridge-line: two rounded peaks sloping off to the south and a range of low, curving hilltops.

I keep stopping to check for ticks. Too soon, I find one crawling up through the wool of my sock. It is round and flattened, like a drop of dried blood with eight stiff, hooked legs. I am going to kill it (crushing it between two stones) but with mixed feelings, because, setting aside its blood-sucking, disease-carrying lifestyle, the tick has something that any human being would envy: some kind of speed-control mechanism for its life. If a young tick finds a warm-blooded animal, the tick buries its head under the host's skin, drinks blood until it almost pops, drops off, lays its eggs, and dies. But if no animal host brushes by, the tick can simply wait, hanging by its back legs to a branch over a trail, its front legs groping at air, for as many as eight long years. Imagine an insect that is eight years old. Imagine a mountain that has been eroding for nine million years.

Frank and I want to follow the river another mile up to the head-wall of the valley, but Erin and Jonathan have stopped to fish for the red-banded trout, a relic of the Ice Ages, a companion of mastodons and saber-toothed tigers, trapped in the Little Stoney Gorge by receding water. So we leave the kids with our packs along the river and hike on without them. Each curve in the trail unveils a new wall as we walk and climb until finally we top a crest at the end of the valley, and all the head-wall circles steep and green around us. Snowfields hang in the tops of the side valleys, where the bare limbs of the aspens catch the light and cast long shadows on the ice. We sit on rough lichens to catch our breath, alone in the Ice Age.

The air seems still, frozen in time, so I am surprised and utterly unprepared—without my pack, without my children—when thunderheads slide over the top of the head-wall. Frank and I turn and hurry down the valley. The river rises as we jog alongside it: heavy rain on the mountaintop. Our children are not where we left them. We jog faster, but the rain catches us, hitting us hard on the backs of our legs. When we finally find them, Erin and Jonathan are sitting on the trail in deep grass on an open slope, counting off seconds after each flash of lightning.

Sound travels 1,100 feet per second, so thunder takes about five seconds to travel a mile. By their count, the lightning is two and a half seconds away and coming fast. We pull on raincoats, debate in high-pitched voices about the safest place in a lightning storm, and run together into a grove of willows. Perched on the exposed roots to escape water that rises in the thick moss under our feet, we raise our hoods and turn our backs to the wind. Geologic time runs a torrent. The mountain picks up speed and races through a thousand rivulets down to the plain that is heaving like a rain-flattened sea. I hold very still, hunching my shoulders to keep my raincoat away from my back, wrapping my arms around my trunk and tucking my hands in my armpits. Rain sheets down my raincoat and drives in through the seams. Thunder slams from one side of the canyon to the other. Thunder on thunder on thunder.

Standing in the rain, we talk about aspens and mastodons and the speed of light, about how trout can hold against the currents of a river in flood. Trout find eddies behind rocks where the current careers around and heads back upstream. So it's the deflected power of the river that holds fish in still water while all the flood rushes past. I suggest that those trout might be a Lesson in Life, but the kids laugh, spilling rainwater from pools on their hoods.

The Smohalla River (in the Sun)

Frank nosed the raft into a reedbed along the shore and Jonathan stepped out into ankle-deep water, holding the bowline in one hand. The raft swung around in the current, levered in parallel to shore and, as Jonathan let out on the rope, dropped into a small bay where it rocked in quiet water. From the shoreline, we walked a faint trail across a stone beach, up the bank between two cottonwoods and into a grassy opening in a pine forest.

We considered the view. A person sitting in the pine opening could see across the river to basalt ramparts at the knees of Bear Mountain. Drinking water? Just down the beach, Alder Creek filtered through a thicket of willows, promising fresh water. One last test. Blocking the sun with outstretched hands, we checked the direction of light to be sure that the opening would have afternoon shade and morning sun.

Yes, yes, and yes again.

So, *Welcome home,* we said to each other, and set to work. Setting up camp is a dance of changing partners and sure steps, played out in primary colors and carefully crafted gear. Jonathan tied the bowline to a thick root. Frank unstrapped the boat bags. Erin and I stepped into the water. Then, pivoting like dancers in a Virginia reel, we passed the gear out of the boat—Frank to Erin to me to Jonathan, who piled it on the shore.

Against gray rock, the heap of equipment shone bright as peeled crayons. A blue coil of rope. A white cowboy hat. A palette of food buckets color-coded to the sky—blue bucket for breakfasts, yellow bucket for lunches, orange bucket for suppers. A frog green rubber bag for the sleeping bags, a blue bag for the tent, four school bus yellow clothes bags, and an orange insulated cooler.

Soon home unfolded, expanded, inflated, until it filled the grassy space under the pines. We popped a big tent up on the grass. Inside the tent, four sleeping bags spread out over four sleeping pads. A clothesline flapping with T-shirts spanned the distance between two trees. On the beach, a kayak and a float tube grew fat with air, fly rods linked to great length, long lines uncoiled. From the supper bucket came peanuts and bourbon. A green stove unfolded from a package the size of a lunchbox. On the stove, rice plumped up white and light in a covered pot that let off little puffs of steam, and, mixed with boiling water, an envelope of pink powder swelled into Texas Chili.

The next morning, we broke camp, stuffing everything back into its container. Then we were a blue rubber raft with a red bottom, 14 feet by 5 feet, self-contained, self-sufficient for four more riverdays and nights.

I have a theory about wilderness excursions like this one. I think that one of the greatest pleasures people get from traveling into a wild place is the feeling of security and safety the experience provides. That feeling comes from knowing that no matter what happens, you can be safe and cozy with only what you have in your backpack or on your raft or trailing behind your skis on a sled. Let it rain. Let it snow. Let lightning strike all around. Let a flood destroy the bridge at the trailhead. From your pack you can pull what you need to make a safe little hearth and home for you and the people you love. Domesticity is the central pleasure of the wilderness experience.

I acknowledge that my wilderness experiences are filtered through a woman's eyes, and that women are supposed to be domestic—two facts that may explain my point of view. But I don't think it is only women who feel the deep satisfaction of playing house in the woods. Thoreau had his cabin at Walden Pond. Henry Beston had his outermost house, Wendell Berry, his long-legged house. Domesticity is everywhere you look in nature writing and wilderness journals, regardless of the author's gender.

Take Colin Fletcher, for example, who in his 1968 book *The Man Who Walked Through Time* describes his camp routine again and again, never leaving out a single loving detail. First he chooses a site with a stupendous view over the Grand Canyon. Then he wedges his walking staff against a rock. He braces his pack against his staff in such a way that the pack becomes a backrest. His sleeping bag lies against the pack. His canteen goes just here. His binoculars go just there. All his equipment is arranged to give him a comfortable, secure home on the slick-rock edge of an abyss.

Margaret Murie, backpacking in the Brooks Range, hurried through a description of the landscape, so she could tell about the bed Olaus made. *Olaus had found a little mossy spot where we could camp. . . . A cold wind had started to blow, but the campsite was sheltered, and somehow as soon as a fire is burning, life always seems possible, even joyous. . . . Olaus gathered some dry moss from the rocks above us and made a fine bed, building up the edges with rocks so we wouldn't roll off. . . . [The next morning,] we started off at eight o'clock, taking a last grateful look at our little haven and wondering what a grizzly might think of that mossy bed!* In the midst of the wildest country left on earth, the subarctic mountains, a scene of the most tender domesticity holds our attention and resonates with our own sense of pleasure.

So deep is the pleasure of keeping house in the woods that when there is no wild land around, people make do with a near substitute—the branches of a tree. When Jonathan was ten, he hauled the top of an old picnic table up into the crotch of a big oak growing at the corner of our city lot. With long spikes, he nailed it in place. He made a ladder by cutting notches into a sapling with an axe and bracing the stair-stepped tree up into the oak. Then he roofed his house with sticks that he had collected one by one and hauled into the treehouse with as much care as a mother robin building a nest. To make a hoist, he attached a rope to a turkey roaster and slung the rope over a tall branch. In a fork in the tree, a pile of pinecones was carefully cached away as defense against intruders. When Jonathan sat in his tree, under his roof, and ate a cupcake pulled up in his turkey roaster, his happiness was complete.

My own treehouse was in a willow in the backyard at 80 Kraft Street. It was a cube made out of discarded window screens, so it let in the breeze and kept out the Ohio bugs. What made it distinctive was that its entryway was low and small, so all visitors had to crawl in on their hands and knees. The entryway was carpeted with a rag rug, and embedded in the rug were thumbtacks, point up, to keep out the boys. If a person didn't know where to crawl, he would grind a tack into his knee.

Jonathan and his friends spent the night in his treehouse. They didn't take much equipment—just sleeping bags and more nacho-flavored corn chips than an adult could eat in a lifetime. At about that age, I slept out in my treehouse, too, with a pile of Nancy Drew mysteries, a box of saltines, grape Kool-aid in a plastic pitcher, a flashlight, a bedroll, and a pillow. Of course a boy did come, and he did get a thumbtack in his knee, which made the whole experience a great triumph.

I think of those times when I walk by the nest of the dusky-footed wood rats down by the river. Dusky-footed wood rats are long-tailed rodents about the size and color of a baked potato. Their stomachs are creamy white, though, and their feet are as dark as if they had walked in walnut husks after a rain. The rats have built a nest in the middle of a mean, thick blackberry bramble, a tangle of canes as thick as thumbs growing ten feet tall and curving down to the ground. The nest is made of five-inch lengths of blackberry canes piled in what looks like a careless heap, except that this heap is higher than my head. Each summer, the blackberry canes grow up over the house and each fall the thicket is stripped, and the house is a foot taller. Anything coming into this treehouse has to find its way over recurving spines a half-inch long. Inside there must be tunnels cut through the canes and there must be nest cavities lined with thistle-silk to shelter the naked, nursing rat babies, safe and warm, but there is no way to tell without destroying the nest.

Another dusky-footed wood rat nest is up in an ash tree, closer to the river, high and fancy like the Swiss Family Robinson's treehouse, which had a balcony and an organ. Sitting in an ordinary house in town, it's hard not to envy the rats or the

Robinsons or Robinson Crusoe whose house was built into the hillside or Anasazi Indians, mysteriously vanished from romantic, magical stone houses built in caves on cliffs. *Gilligan's Island* spoke to the same fantasies. A boy I met in college told me his family owned an island with white pines and blueberries and three wooden cabins; for many years after we were married, we slept as often as we could in Cabin Two, which had no electricity but was close enough to the water to let the soft lapping lake speak through the floorboards. When Frank and I had children, we read them Jean George's book, *My Side of the Mountain,* again and again, relishing the story of a boy who lived happily in a burnt-out stump, even in the winter when it snowed, alone except for a peregrine falcon.

Late one winter, many years ago, we skied with our children into Fawn Lake in the Cascades, pulling our camping gear in sleds harnessed to long poles at our waists. Gray moisture was seeping down through the hemlocks, and it seemed to be getting dark, even though it was the middle of the day. All of us were uneasy, playing out a plan that was so obviously a mistake in such miserable weather, but our children were especially unhappy. Slouched over ski poles braced against their armpits, they stood on their skis with their arms dangling, disgusted at the prospect of spending the night on this soggy, cold, monochromatic mountain. Frank and I left them at the shore of the lake and skied around in the woods, trying to find a flat place to pitch a tent. By the time we returned, the children had each collected an armload of pinecones, boughs, and fallen sticks and piled it at the base of a lodgepole pine to make a seat. They had collected another heap of sticks for a footrest. Little skis were stuck at odd angles into a snowbank, and wet mittens hung from the branches of a nearby tree. There they sat on their little dry seats, leaning against their tree, passing a bag of pretzels back and forth through the fog.

Human beings have simple needs—to eat, to drink, to nest, to have children and keep them safe. We play out the satisfaction of those needs with complex equipment and long excursions to wild places, where we spend the end of each day happily engaged in home-making. The greater the challenge, the greater the comfort when we overcome that challenge. Violent weather, isolation, extremes—all these raise the value of warmth and safety, and so it is necessary to seek out storms.

I wouldn't argue that this is the only reason people go to wild places, but I believe it is an important one. *How far can I go . . . ? How high can I climb . . . ? How little can I bring along . . . ?* are questions that end in a phrase that is usually unstated, but always understood: *and still be safe and warm?*

The McKenzie River

We advanced along the McKenzie River Trail with the effciency of inventory clerks. Brown to purple flowers, heart-shaped leaves: *wild ginger.* Leaning, leafy stems, dense cluster of white flowers: *false Solomon's seal.* Three broad fan-shaped leaflets: *vanilla leaf.* The deer fern we knew, but a pale tattery fern stymied us. *Plants of the Pacific Northwest* said it might be a lady fern, easily confused with wood fern, page 26. Page 26 said it may be a wood fern, but check for sori that are crescent-shaped, rather than round. We trooped back down the trail to check the sori. There were none: a defective fern.

Single bell-like flower under each leaf: *twisted stalk.* Heart-shaped flowers in nodding clusters: *Pacific bleedinghearts. Maiden-hair fern. Monkey flower.*

I learned most of these names walking forest trails with my father, who was a taxonomist. *Taxonomy,* from the Greek word *nomos,* a rational principle, + *taxare,* to rate, to place a value on, to assess, to call to account. In the field, my father carried a metal cylinder on a strap hung over his shoulder. A narrow, hinged door opened along its length. Whenever he found a plant that interested him, he cut it off at the soil, laid it carefully inside the cylinder, and shut the little door. At home, he spread newspaper on the kitchen table and laid out all the plants side by side. With a hand lens and a pair of tweezers, he dismembered each flower, separating all the petals, pulling open the tiny ovaries, following the paired steps in a dichotomous key. Once he had identified a plant, he wrote genus, species, date, and location on a sheet of newspaper and pulled the newspaper up over the face of the flower. By the end of the evening, he had a stack on top of a wooden pallet: cardboard, blotting paper, newspaper, specimen, newspaper, blotting paper, cardboard. He topped the stack with another wooden pallet and, kneeling on the press, cinched it tight with two straps. After many years, my father had taught himself the name of every plant in the weed fields and beech-maple woods around our town.

When our children came into the world, we taught them the names of things. Richard Scarry's *Best Word Book Ever* was the first bible in our house. We sat four abreast on the couch with the big book spread across our laps, and we pointed at the pictures and said the names. *Bus, Car, Bicycle, Truck. Pencil, Crayon, Paper, Book.* Jonathan's eyes followed Erin's little pointer finger up and down, across the page, through the book, as she pronounced all the words in the world. At that time, teaching them words seemed like the most important thing parents could do for their children.

When Lewis and Clark came into the West, they had the job of naming everything they saw. They named the trees: For the lodgepole pine, they chose *Pinus contorta,* because the lodgepoles they saw were coastal trees, twisted and dwarfed by Pacific Ocean storms. They named the birds: Clark's nutcracker. Lewis' woodpecker. And later, scientists named wildflowers in their honor: for Meriwether Lewis, a pink flower that blooms close to the ground in high, dry desert gravel—*Lewisia rediviva,* Lewis come-back-to-life, bitterroot; for Captain Clark, farewell-to-spring, *Clarkia amoena,* an evening primrose, a fan of pink petals, each petal with a purple spot in the center. Lewis and Clark named each river as they floated downstream: "Passed a small creek on Stard. at the entrance of which Reubin Fields killed a large Panther. we called the creek after that animal Panther Creek. The day before, found the river much crouded with islands both large and small and passed a small creek on Stard. side which we called birth Creek," because it was Captain Clark's birthday.

We have played the same game on our own river trips, landing on a gravel bar thick with willows and piles of mud-washed logs. We name the high places first, and then the low, hidden places. And then we conduct each other on guided tours, telling the names. By this ritual, the island belongs to us. We own Anniversary Island with Picnic Point and Dead Duck Beach. We own Family Island, canoeing out to reclaim it every summer when it emerges from spring floods, with new bays and logjams in need of names. We own our farm, not because of the mortgage, but because we have walked the land together and named the Oak Knoll, the Ash Swale, Turtle Point, Winter Creek, the North Hill pasture, and, above the railroad tracks, the Magic Forest. My father named the Magic Forest on his last visit west, when even his hand lens couldn't bring flowers into focus, and the plants confused him, so far from home.

Children bring their dolls to life by giving them names. Names transform animals into family members. In some religions, nobody can have eternal life—not even tiny babies— until they have been baptized, given a name. In a single word—*ilira*—the Inuit people bring to tangible life the awe and fear that possess them when they see a polar bear approaching across the ice; in another word—*kappia*—they name the apprehension that seizes them when they cross thin sea ice. In Genesis, all the parts of the universe are drawn out of fluid chaos by their names. "God called the dry land Earth, and the waters that were gathered together he called Seas." To be is to be named.

Yet no one is allowed to know God's name. He is called *yhwh,* a word without vowels, an obvious fake, a name that does not name. It is an issue, clearly, of disproportionate power. The power to name is the power to create, and the power to destroy.

Truck drivers call their CB nicknames *handles,* and this seems right. A handle gives leverage. Last year, a police officer leaned against the driver's side of my van. He wore reflective glasses that revealed my face, not his. "May I see some identification, please?" I surrendered my driver's license, giving my identity over to the policeman. On the other side of the Earth, Soviet citizens used to tell each other this joke: "I will describe to you the happiest day of my life. A knock came at the door of my apartment, and I opened the door to two men in trench coats. 'Is your name Ivano?' one said. 'No,' I said, 'he lives upstairs.' "

On duff under an uprooted fir, we found a slime mold, a bright yellow hauck of phlegm. When the kids poked at it with pine needles, every part they poked turned to water and disappeared. The slime mold defies classification and so defies naming. It isn't a plant because it moves around under its own power. And it isn't exactly an animal. After trying for decades to pin it down, taxonomists put the slime mold in its own kingdom—*Protista,* none of the above.

I wonder what it would be like to go into a forest where nothing had a name. If there were no word for tree stumps, would they sink into the duff? It's possible: After all, earth without form is void. And if we started over, giving names, would any fact about the forest compel us to name the same units? Would we label trees? Or instead would we find a name for the unity of roots and soil and microorganisms? Or would we label only the gloss of light on leaves and the shapes of shadows on the bark? How would we act in a forest if there were no names for anything smaller than an ecosystem? How could we walk, if there were no way to talk about anything larger than a cell?

All along the McKenzie River Trail, there must be things we do not see, because they have no names. If we knew a word for the dark spaces between pebbles on the river bottom, if we had a name for the nests of dried grass deposited by floods high in riverside trees, if there were a word apiece for the smell of pines in the sunshine and in the shadows, we would walk a different trail.

The Metolius

On one of the dark, shiny, rain-slick nights that we get in Oregon, I ran over a possum with my truck. I saw it just at the edge of my vision, jerked the wheel to get out of the way, and hit it dead on. Dead possums are about as common as stop signs along the roads in my town, but this was my roadkill possum, and to make it all unspeakably worse, a baby no bigger than the palm of my hand was nosing around the body that lay open and steaming across the fogline. I gathered the baby up in my cupped hands and took her home to raise as if she were my own child.

At first I fed her warm milk mixed with Karo syrup, squirting it into her mouth with an eyedropper every four hours all through the night. I stroked her hindquarters with a warm, wet cloth to make her defecate, washed her tail in soapy water, and carried her with me, clean and sweet in her refuge under the hair at the back of my neck. When I thought she was tired, I put her to bed in a box of rags and watched her doze, a naked tail encircling her body, milk dried around the edges of her possum grin.

After I had weaned her to cat food and taught her to drink from a bowl, the time came for the ceremony of healing, the time when she could be returned to the sunny thicket where she was born. Brimming with the damp hope and sorrow of a mother on the first day of kindergarten, I lifted her out of her box of rags and set her in warm grass under blackberry canes tangled in an ancient tree. Faster than I could lift my eyes, she charged up my arm, jumped onto my neck, and sank her teeth into my throat. When I grabbed at her with both hands and yanked her away, her teeth raked through my skin. I held her away from my body to keep her from biting again, but she drooped in my hands, limp, insensible, as if she were dead.

I swiped at the blood with my T-shirt. Stupid to think I ever stood a chance of undoing what I did to the possum mother, to think I could find redemption in good works. I held the heel of my hand hard against my neck. Stupid to think that one mother is as good as any other. The hurt was sharp, as if there were dirt or acid on the possum's teeth. How can I claim to be a mother, when I slap away a baby who comes to me for protection? The baby knew what to do in a strange place, in the terrifying glare of full sun. She scrambled for the thick fur on her mother's neck and held on for dear life. But there was no possum mother, and there was no fur.

In the end, I left the possum in an abandoned barn with a bowl of water and a week's supply of cat food. That's all I know about what happened.

Last spring, I saw a student on campus in a T-shirt that said, "Love Your Mother," and showed an astronaut's photo of the cloud-splashed Earth. To be honest, the T-shirt irritated me, and so did the student. I wondered if he ever wrote to his mother or remembered her birthday. I wondered if the student had figured out what it means to think of the Earth as a mother: Mother Earth, Magna Mater. I wondered if he had even thought about it, if he had any idea how complicated the analogy is, if he was ready to take a stand as to exactly how—in what particular way—we are supposed to understand the connection between the Earth and a mother, and what difference that connection makes for the way we live, what obligations it imposes.

Maybe he's pointing out that the Earth, like a mother, is productive, reproductive, and that we humans are born of the earth, just as the Earth is born of star-matter spinning in long spirals from the explosion of the Big Bang. But what difference does that relationship make? Humans may be made of earth-matter, but so are Oldsmobiles, and that doesn't make the Earth into the Mother of Oldsmobiles, and it doesn't explain the source of our obligations.

Sometimes I think the earth-mother analogy is just wishful thinking. It's comfortable to have a mother around the house. Mothers can usually be counted on to clean up after their children. Mothers provide for their children; they cook regular meals and pack healthy snacks. Mothers are warm-hearted and forgiving; they are the ones who follow crying children to their rooms and stroke their hair, even if the child's sorrow is shame at his treatment of his mother.

What great good fortune it would be for us if the Earth were a mother like this. If people are going to bury metal cylinders of strontium-90 inside sacred desert mountains, if we are going to poison coyotes and hang their corpses on barbed-wire fences, if we are going to send sulfur into the sky until the needles fall off pine trees, if we are going to poison the water we drink and the fruit we eat, then I suppose it is nice to think that the Earth is a mother who will come behind us and clean up the mess and protect us from our mistakes, and then forgive us for the monstrous betrayal.

Or is the college boy's T-shirt really saying something important about our obligations, telling us to treat the Earth as lovingly as we would treat our own mothers? The thought made me uneasy; I wanted to run after the student and warn him to be careful, to learn more about history, to study Europa or the Sabine women, to think about how he treats his own mother, before he wears a shirt that gives such reckless advice.

Analogies, once drawn, are powerful, magical, impossible to control. This one cuts in ambiguous and dangerous ways.

When I was a teenager, my mother's dying was part of everyday life. I didn't understand why she had been singled out to die, but the scientific explanation was clear enough: Bronchiectasis was destroying her pulmonary cilia, so her lungs couldn't clean themselves. Over time, they would get dirtier and dirtier, overflowing with dead cells and phlegm and the gritty buildup of tar from the cigarettes my father used to smoke, and she would not be able to breathe any more. We learned this. And then we tried not to think about it, tried not to let it make a difference in the way we lived.

I can remember the steel feeling in my own lungs each time I woke in the dark and heard my father's step on the stairs as he came up to tell my older sister that he was taking our mother to the hospital again, that he would be back as soon as he could, that she should make everybody help. Then I remember lying in the dark bedroom, listening to the frozen gravel crunch under the car backing out of the driveway, listening to my sister's jagged breathing, wondering how children could continue to live after their mother died.

While my mother was in the hospital, my father insisted on constructive responses. "It's not a problem for me, really," he would always tell the neighbors when they offered to help. "I have three other little women in the house, and I will get along fine. Nancy, the organized one, can keep the house clean. Kathy, the hungry one, can cook. And Sally, the cheerful one, can make us laugh." So we cleaned and cooked and laughed, cleaned and cooked and laughed, while the airborne detritus of ordinary life hardened in my mother's chest.

We tried as a matter of policy to postpone all crying until after she had died, because every time any of us cried, she held us close and sobbed herself, and choked on the mold in her lungs and started to cough. We would pull away and stare at her as she struggled, terrified that we had killed her, killed her by not being strong enough. Because we knew that sorrow was too dangerous to indulge, we forced our mouths to tell a funny story—a kid fell off his seat in algebra, a bird flew in the window during second period—and after a while it got easy just to tell the stories and never say what needed to be said.

I was away at college when my mother died, so I don't know what the end was like, but I have always imagined my mother retreating slowly into a private world as cloud-wrapped as Venus—perpetual deep twilight, red-tinged at the edges, steamy with carbon dioxide, the hot, still air too thick to breathe. But my father says it wasn't like that. All the week before she died, he says, her hands were cold. Each morning, he would put another blanket on her bed, trying to take off the chill, and each evening he would read aloud from Thoreau's journals. She died when they were deep into the days of November, "a still, cold night. The ground is frozen and echoes to my tread. So far have we got toward winter."

Now I look back in grief at events beyond my power to repair. When I think of the Earth, when I think of my mother, desperately damaged and then wronged again by silence, it is the power of mothers to forgive that holds my attention, and the terrible questions of redemption and regret.

When the Earth is whole, it is forgiving, resilient, covering burned-over land with blankets of fireweed, then alder thickets, then cold pine forests; filtering water through wetlands bright with sky; in every sun-washed cell, creating oxygen; again and again transforming death into life. But once it is damaged, profaned, the power of the Earth to heal itself seeps away, and with it goes its power to nurture us. In a weakened world, we children of the earth may turn against the land, cut the last steep slopes for firewood, pour chemical fertilizers onto worn-out fields, sanitize wastewater with poisons, spray more pesticides, dam more rivers, burn more oil, bear more children—not out of malice, but out of fear of the future in a world that may not have the strength to help us. We turn our faces away, willfully separating ourselves from the earth, never acknowledging that there may be no chance for healing, refusing to remember what we have done and what we have failed to do. Then, who can forgive us?

In a ponderosa pine forest just north of Santiam Pass in the Cascade Mountains, the Metolius River flows full-bodied out of a crack in a hillside. If you stand near the base of Black Butte, you can watch the river's birth—50,000 gallons of ice-cold water pouring out each minute, an entire newborn river surging from the boulders into the forest and flowing north along the base of the high peaks, around the back of Green Ridge, and down through a canyon five hundred feet deep. For twelve more miles, the river flows between basalt bluffs and then it stops dead, dammed into an impoundment enclosed by raw cliffs: the Bureau of Land Management's Lake Billy Chinook.

I have stood under the bluff at the south end of the reservoir and looked back up the inlet where the Metolius River comes to its end. Through the sweet mist of mockorange, under a canopy of lodgepole pines, between the roots of willows, the river slides clear and cold down stair-stepped boulders to the impoundment. At the base of the canyon, the river plows into gray water, slows to a stop, clouds at the edges, and drowns in the warm and silt-choked reservoir.

A few years after my mother died, I brought my father to the headwaters of the Metolius. It was his idea to come on a winter night, timing the trip so that the full moon would light the shining river and the broad face of Mt. Jefferson across the meadow. The path to the river was treacherous with patchy ice, so I picked my steps carefully, supporting my father by the arm, warning him about ice under his cane, terrified that he would fall and break a disease-brittled bone.

When we came to the edge of the water, a heron jumped into the air, lifted itself with one beat of its heavy wings, and disappeared. Scalloped clouds blew across the face of the moon, their edges glowing, then dissolving back into the dark. Barely visible in the shadow of the hill, boulders spread apart and the river rushed out, full and frothing, alive with oxygen, pouring downhill as if it could flow forever. I helped my father brace himself against a tree and slowly scrape the snow off a bench. We sat side by side, looking into the darkness, pulling cold, pine-drenched air deep into our lungs, saying nothing.

Three

Bear Creek

For forty miles along the Pacific coast, sand dunes pile up like
surf in a dazzling, blowing sea. There are islands in the sand
dunes, small round pockets of forest. The dunes break against
their shores, and fingers of sand wash in among the tree trunks.
At the southern end of the dunes, Bear Creek cuts through from
east to west. Where the roots of salal and beach pine settle the
sand, the creek runs a predictable course between steep-sided
banks. But where the sand has its way, the streambed takes a
winding, indirect path—hesitant, haphazard, seeking the low-
est ground, shifting course from year to year, seeming not to care
if it gets to the sea or not; so the sea comes after the creek, breaking through the foredune and chasing up the deflation plain
until it touches the creek, grips it by its narrow shoulders, and
pulls it into the ocean. In the heart of the dunes, there are bears
and cougars. Or at least, I should say, I have seen their tracks.

It's hard to get into the dunes, since they are protected by a
tangle of swamps, sunken forests, and high hills so thickly
forested with salal, salmonberry, swordferns, and vine maples
that nothing but small animals travel there—white-footed deer

mice, maybe, or vagrant shrews, to judge from their tracks. It's possible to come in by way of Bear Creek, shoving a canoe down stickery banks and floating around the slow curves. But you have to be prepared to get out of the canoe, if you need to, and drag it over a pile of logs drifted in on high tide. Some of the logs will be floating, but you won't know this until you step on one, by which time it will be too late. You will slowly roll into water thigh deep. If you want to hike into the dunes, the most direct route climbs through dense thickets of Scotch broom. In the springtime, you will be picking your way through groves of yellow flowers that smell like lilies, following trails cut by little coastal deer that reach only four feet at the shoulders. The easiest way to walk into the dunes is the longest way, down the beach and along the coastline to Bear Creek, then up through the streambed, wading in shallow water. Quicksand is the uncertainty here, soft spots where springs bubble up through sand on the bottom of the river, making the sand too liquid to hold any weight.

But once you reach the bare sand hills, the landscape opens up in front of you—shining sand hills all the way to the gleaming line that marks the edge of the sea—and the way is clear to walk along the ridgelines to the heart of the dunes.

On a calm day, it's tempting to think of sand dunes as a thing, a place, an entity that persists, like any well-behaved geographical location, from one season to the next. But the sand dunes aren't always like this. In a full gale, the dunes become liquid. Sand pours up the windward side of the dune and blasts across the ridgeline in explosions of light, curling in breakers that tear apart in the wind. Sparkling sand falls down the leeward side and lands in soft shadows. Solid earth blends seamlessly into sunlight. The air hisses; under that, a deeper sound, a roar like surf on rocks. If you are in the way, the wind will blow your hair back, scour the skin on your windward side, and fill your shadow with heaps of sand.

When the wind stops, suddenly and without explanation, sand that has flowed as if molten hardens into sweeping arcs, high hills, and deep wells with just a sheen of water at the bottom. The wind has carved valleys and shaped mountains, moved sand hills a mile down the beach. Except where a stick or a bone compressed the sand, the wind has blown the surface off the land, leaving a monument valley of miniature pinnacles and balanced rocks, a world at the dawn of creation.

Early mornings, the sand dunes shine with moisture from the night. The air is so still that it is possible to hear warblers in the forest islands and surf washing onshore several miles away. This is when Frank and the kids and I like to come into the dunes, to see what is afoot. There are stories here, and puzzles, dreams, places no one has ever seen before.

One April, many years ago, we hiked in from the coast side of the dunes and walked barefoot up the center of Bear Creek in water no more than a foot deep. Little sculpins appeared out of nowhere, darting away to bury themselves in puffs of sand. Sandpipers with improbably skinny legs blitzed along the shore, peeping incessantly. We passed a beaten-down path where river otters slid into the stream. And then the bed of the river gave way under my left foot, and I found myself crawling across the bottom of the river toward firmer ground. Frank and the kids laughed until they had to sit down, overcome, but I noticed that after that, they tested their steps, trying to locate quicksand by the subtle quivering of the sand bottom. When they skirted the wobbling quicksand beaches of a little pond, they walked in a half crouch with their arms outstretched and lifted their feet from the knees, like gulls.

For a while, we followed bear tracks that we found next to the river—two sets, one big, one small. The big set paralleled the river in a sequence of regular steps, one foot and then the other, unswerving, serious. The little set wove in and out of the big set, dropping into the river, cleaving blocks off the sandbank, turning back on themselves, wandering up into loose sand, over to a patch of tangled kelp and back to the bank, a full circle around a plywood board, then long loping tracks back to the edge of the river. Frank and I looked at each other and laughed, but Erin and Jonathan didn't see anything funny, and besides, they were already climbing up a steep sandbank to scan the horizon for bears. When there were no wild creatures to be seen, they lay down on the crest of the dune. Tucking their arms close to their sides, they closed their eyes, held their breath, and let themselves roll all the way to the bottom, turning and turning until their pockets filled with sand and they slowed to a stop in the deflation plain.

We set up camp in a secret spot we had found just by chance, a long time ago. The campsite is a little beach enclosed by a curve of the creek, deep in the tallest dunes, far enough from the ocean to be protected from the wind but close enough that, morning and evening, air moves up the river from the sea, bringing a salty, low-tide smell and a low braying that we think must be the foghorn at the mouth of Cow Slough Bay. An ancient log runs the length of the beach, battered to splinters, polished silver, pushed upstream and stranded by a storm. The campsite backs up against a forest, and the forest backs up against a bare dune five hundred feet high.

We camped there for three days and three nights. On the morning of the third day, Erin and Jonathan had gone up the beach to gather driftwood for a breakfast fire. Frank and I had just poured another cup of coffee and were sitting on the log to watch the kids. Wild splashing downstream brought us to our feet. A fawn ran hard up the river, lifting its hooves high, raising sheets of water into the fog. Passing within a few feet of us, it bounded up the bank. The fawn slowed its pace and walked delicately and deliberately toward Erin. It was just a baby, with enormous ears and eyes, spindly legs, and a wash of spots across its back. A few steps more, a few steps more, and the fawn got so close that Erin could have reached out and stroked its neck. When it moved closer still, Erin hesitated, then backed away. The fawn raised its head, took another step toward her. Erin stepped back, stepped back again. The fawn moved with her, not two feet away, looking into her eyes. Completely unnerved, Erin broke and ran for camp. The fawn backed into the salal and disappeared.

None of us knew what to make of it. We checked our memories against each other's. Did you see what I saw? Just that close? A fawn? Searching for clues, we walked back downstream, retracing the fawn's path along a line of heart-shaped hoofprints that cut, splayed and sharp-edged, deep in the sand. Around a bend on the far side of camp, we came across the broad running tracks of a coyote superimposed on the fawn's little prints—wide-spread at first, then closer together, slowing. Nearing camp, the coyote's tracks stopped and paced, turned, and sprinted back into the fog.

We caught crayfish that afternoon and put them in a pen made of sand at the edge of the creek, but they had all escaped by nightfall.

Each evening that we camped in the secret place, we climbed to the top of the highest dune, going early to get a good seat for the sunset, knowing full well that we were the only people for five thousand miles to the west, and the only people for several miles in every other direction. Getting to the top of the dune was a heart-pounding climb through soft sand that slid down the hill at a pace just a little bit slower than we could scramble up. On the way back to camp in the dark, we would run toward the edge of the dune and jump into space, assuming that the sand would catch us softly and carry us downhill, depositing us just at the edge of the river. But until then, we sat in a small grove of pines and beach grass at the top of the dune, facing out toward a narrow silver strip of ocean and the deep sky.

We couldn't predict what would happen next. The first evening, nothing really happened: The air gradually cooled, seagulls flew north against the wind, the day faded away until it was night, and we walked back to camp. But on the second night, we heard coyotes calling far down the beach. And the third time we sat on the dune at nightfall, the sun dropped fast to the edge of the sea where it exploded bright as a struck match, catching the clouds on fire and burning the waves as black as soot. We waited there on the dune, sitting deep in the beach grass, in the dusk, until nothing was left of the sunset but the faint smell of woodsmoke on damp air.

That night, we slept with the tent zippered up against the cold. When Jonathan got out of the tent in the morning, he found fresh tracks in the sand: four round toes with no sign of any claws, a central pad, the whole track about three inches across, impressed maybe a quarter inch in damp sand. Following the tracks, Jonathan walked from the log to the door of the tent, stopped a moment, then turned to the right and slowly circled the entire tent, then paused again at the door, then walked up the beach to a salal thicket, where the cougar tracks disappeared down a path too small to follow.

Years later, when the kids had grown, Frank and I came back to the hidden place alone. Work crews had cut a road through the forest to the edge of the river and paved the surface with black asphalt. At high tidewater, about where a logjam had been many years before, they had bulldozed out a parking area and built two cedar-paneled outhouses and a three-sided interpretive sign that explained in great detail the action of the dunes. There were sketches illustrating how wind lifts the sand and marches the dunes to leeward, and photographs of the foredune and deflation plain. Drawings showed the animals one might encounter on the dunes and identified the shells that might wash up on the shore. PROTECTED AREA, a sign said. From the interpretive sign, an asphalt trail led out toward the beach, but before it got far, it was swallowed by a great triumphant mound of sand that lifted, sizzling, in the wind and poured over the windward parking spaces, covering the neat white lines and concrete curbs, drifting around the posts that supported the interpretive sign.

The logic of the sand dunes is different from the logic of other landscapes. In a forested valley or an oak savannah, rocks and grasses and woody plants are the *given,* the facts, rooted and present, undeniable, reliable. But on piles of shifting sand, anything is possible, and this is the triumph, the delight, the beauty of the dunes. Walking in sand, my mind runs ahead like a child, making connections, drawing analogies, following uncertain lines of inference, affirming conclusions, leaving an erratic trail of footprints.

The dunes are a landscape of assumptions and evidence and inference. When I walk in the river I have to assume that the sand will hold my weight, and maybe it will and maybe it won't. It takes a certain set of assumptions to walk on a logjam: assumptions and an extra pair of dry pants. On the dunes, animals explain themselves only in the abstract language of signs in the sand. Ravens leave wandering rows of Xs; coyotes, a path of Os. Beetles leave a double track of semicolons. Deer leave small hearts. The sand dunes are a given, but only in this way. They are speculations—tentative, subject to change. In the dunes, visitors have to accept what is given and then see where that takes them, see what follows from that, step by step. Nothing about the dunes can be taken for granted.

Aguajita Wash

Species List
Aguajita Wash, March 22–27

Barrel cactus
Fishhook cactus
Yucca
Saguaro
Palmetto

We walked out of the Phoenix airport into the night and there they were under fluorescent lights—a whole garden of plants we had never seen before. Fleshy cactus plants towered over our heads, looking not only strange, but exaggerated in their strangeness, as if children had drawn them with crayons. Each enormous plant had its own plastic nameplate, stuck into the ground on a metal stake. The kids and I carefully copied their names into a little notebook, while Frank stood in the midst of a pile of luggage, trying to fit all our camping gear into the trunk of an Oldsmobile from Hertz

Century plant
Prickly pear cactus
Pallid bat!

I had seen pallid bats in books, but never before in real life, and I'm not sure I had believed in them any more than I believed in ghosts. But right there in the parking lot, a white bat materialized at the edge of a halogen light, skimmed the cactus, and disappeared into the darkness by the airport security station. I got out my pencil and wrote, "pallid bat," exclamation mark.

Other people keep travel journals, but we keep species lists. Somewhere in the house, we have lists of the plants and animals we saw on each of the trips we have taken. Lists cover the end pages of field guides and show up now and then on pieces of paper folded inside maps and snapped into notebooks. When I read the names of the animals written all across a page in the barely contained energy of a child's handwriting, the sights and sounds, the smells, the amazement of the original experience flood into my mind and make me smile.

Raven
Black-billed magpie

The first night, we stayed in a motel at the end of the west-bound runway and ducked incoming planes in our sleep. The next morning, exhausted, we drove the car down the highway through endless suburbs. Swaying like a boat in a swell, tuning the radio to Mexican music, watching for living things, we headed for the Sonoran Desert.

> Jackrabbit
> Rock squirrel
> Gopher snake (dead on highway)

Each trip we take, we get in an argument about whether or not to list roadkills in the species list. If we don't include them, we lose some of our most spectacular discoveries. But on the other hand, you don't usually see much of a roadkill—just some tufts of hair sticking up, or pink flesh. So we settled on this sad rule: *If you can recognize it, it counts.*

> Honey mesquite trees
> Creosote bushes
> Red-tailed hawk
> Cottonwood trees
> Rabbit brush
> Sagebrush

I had thought this desert would be like the Great Basin deserts we are used to—brown barren places with junipers spaced evenly, arms raised like drowning sailors. So the spring-time Sonoran Desert surprised me. The plants were fierce, like they are supposed to be in a desert: Every plant had spines, poisonous hairs, hooks, claws, or sticking seed pods. But the spines and stems were softly green and so were the little leaves, so the desert landscape rolled away like waves on a pale green sea.

> Mormon tea
> Desert holly

We pulled the car into a berth between a Silver Stream and a Winnebago. It was a good campsite, at the edge of the campground, where coyotes might wander through at night.

> Zebra-tailed lizard
> Western fence lizard
> Scorpion

As soon as Jonathan was out of the car, he was stalking lizards—turning rocks, poking under mesquite trees, picking his way between cactus, getting cholla joints stuck on the toes of his basketball shoes. He found a scorpion under a rock and brought it back to camp, gripping its stinging tail between his thumb and forefinger. We looked it over from every angle, and then we made him put it back.

> Broad-tailed hummingbird
> Cactus wren (with nest)

> Gila woodpecker
> Cottontail rabbit

We no sooner got the charcoal lighted than a desert thunderstorm charged up over the western mountains like a Comanche war party. The kids retreated to the Oldsmobile. Frank and I ran back and forth through torrential rain between tent and hot dogs and car, trying to keep the fire going and supper cooking and the floor of the tent dry. All the while, the charade was closely monitored by white-haired women behind the louvered windows of their motorhomes, women washing dishes with binoculars hanging around suntanned necks.

The storm passed just before sunset. In its wake, pavement steamed in hot light, cactus wrens sang from the tops of the saguaros, and Frank and I opened another beer.

> Great-horned owl (calls only)
> Little brown bat
> Little pocket mouse
> Desert kangaroo rat
> Desert wood rat

After dark, we lay on our stomachs on the picnic table, peering down over the edge. The beams of our flashlights lit cones of light through pools of desert air. First came the pocket mice, looking for scraps of food that might have fallen off the table. Pocket mice are tawny little rodents, their tails long and lush, tipped with a white flash of fur. They flipped their semaphores back and forth while they stuffed food into their mouths with tiny paws. Then came the kangaroo rats—darker, fiercer, as big as fists. One rat sank its front teeth into a weiner and carried it off, dragging it between its legs the way a cheetah carries an antelope.

> Mourning doves
> California quails

The doves started to call before dawn. I stretched out in my sleeping bag in the cool air, listening to the clear drops of dovesound, smiling to myself, knowing that soon the sun would rise dazzling from behind black hills and hit me full in the face.

At breakfast, two forks and a spoon were missing from the picnic table. We blamed the wood rats, of course, and went straight to their disgraceful nest in the next campsite to demand our property. We found an unruly pile of cholla joints and half-burned tinfoil stuck between the paddle blades of a prickly pear, but no silverware. Later, Erin found the forks in the backseat of the Oldsmobile.

Opening our backpacks in the morning sun, we divided up equipment: flashlight, binoculars, toothbrush, pocketknife, sleeping bag, underwear, socks, T-shirt, wind parka, meals, and three quarts of water and a bag of jelly beans per person. Frank carried the compass and the tent. We loaded the backpacks into the Oldsmobile and headed for the backcountry.

> Scarlet gilia (blooming)
> Anna's hummingbird
> Desert willow
> Garter snake
> Curve-billed thrasher
> Mormon crickets

Thirty miles from camp, twenty miles down a dirt track, we left the car and set out to follow the Aguajita Wash through the desert. This was no bare expanse of sand, but a dry creek-bed carved through rock standing in jagged hills and lumpy monoliths.

> Mexican jay
> Vermillion flycatcher
> Side-blotched lizard

Only a few steps from the car, Erin turned over a rock and found a rough-edged lizard the color of stone. She pounced, trapping it under cupped hands. She lifted it to eye level. On its stomach was a patch of color as brilliant and glittering as the blue green of a tropical sea. The lizard struggled free and leapt to the ground, leaving behind its tail. The tail arched, snapping up into the air, somersaulting. Erin dropped it like a lit match. It flipped and popped across the gravel, jumped a rock, and lay twitching on the hot sand.

> Desert iguana
> Whip-tailed lizard
> Cat's claw tree

The Spanish word for a dry riverbed like the Aguajita Wash is *arroyo*. My sisters and I used to be terrified of arroyos, because even though we grew up in a northern city where the only cows were those printed on the milk jugs, our mother sang us to sleep with the "Cowboy Lullaby," a song that cowboys sang to quiet cattle at night. It has a verse that goes like this:

> *Nothing out there on the plain that you folks need.*
> *Nothing there that seems to catch the eye.*
> *Still you have to watch them or they'll all stampede,*
> *Plungin' down some arroyo bank to di-e-e.*

Back then, none of us knew what an *arroyo bank* was, but we knew it meant death.

Now I know that arroyos are magical places: dry riverbeds carved into the desert floor, concentrators of shade and water, and thus concentrators of life; paths lined with burrows and scurrying with lizards and snakes, buzzing with bees; a hangout for hummingbirds and dazzling bright wildflowers; paths wandering, branching, disappearing, dropping into dank canyons, rising onto sand banks blinding with sun, filling to the banks with light and birdsong. Sometimes dry waterfalls shade pockets of water that vibrate with mosquito larvae tangling in garish green threads of algae. Heart-shaped hoofprints of peccaries mark the shoreline.

We wandered down the dry wash, radiating heat from our faces, looking in all directions, chasing after lizards, stopping to watch hummingbirds, eating jelly beans.

> Bur sage
> Collared lizard
> Smoke tree
> Wild cucumber
> Manzanita bush
> Western kingbird
> Palo verde thickets
> Gila monster

From the time our plane landed in Phoenix, we had been looking for Gila monsters—rare and elusive lizards, paleolithic, poisonous. But as it turned out, we came upon a Gila monster when we were looking, rather desperately, for our packs. We had cached them in a pile of rocks while we went exploring. Knowing the dangers of being without water (and the dollar value of our equipment), we each had located the cache against landmarks, using the compass. Unfortunately, the packs were not where we reckoned they would be, certainly not "on a line 5 degrees off south of the top of Sheep Mountain," which was the bearing I had committed to memory.

So we each went to where we thought the packs might be and, calling to each other so we didn't lose a child too, we wandered around until I circled a creosote bush on a sand flat and encountered a Gila monster. It looked just like its picture in the field guide, only much, much bigger: black with yellow spots, skin lumpy as if embroidered with Indian beads, forked black tongue flickering like a snake, overstuffed, fat legs sticking out to the side, and big—easily two feet long. I yelled and everybody came running to see this legendary lizard.

> Desert globemallow
> Rock nettle
> Turkey vultures

Finally Frank went back to his compass and his bearing and found the packs by walking a line due west of the hole in the mountain, which was where they were supposed to be all along. We set up camp in a little sandy cove sheltered by a circle of rocks. To one side, under a desert holly, Erin found an ancient weathered piece of pottery, brown on one side, black on the other, with the marks of smoothing thumbs still visible on one edge. We had done our best to lose ourselves in the wilderness and so we had come to a place that had been the center of someone else's life. When the wind blew that night, the spines on the saguaros whistled like an old man breathing.

> Canyon wren
> Sidewinder rattlesnake
> Brittlebrush

That night we went out after sidewinders. Lots of people think that the desert is full of poisonous snakes waiting to chase you down and bite you, but the fact is that you usually have to work hard to get within biting range of a poisonous snake, especially a nocturnal snake like the sidewinder.

With flashlights and canteens, we set off for the sand dunes as soon as it was dark. Right away we found the strange tracks of the snake that moves by throwing a loop in its body and then pulling the rest of itself over to the place the loop lands, leaving a track of parallel foot-long lines in the sand.

It wasn't long before Frank had a sidewinder pinned in the beam of his flashlight. The snake was small—maybe 18 inches—and pale as a beach in moonlight, rough-edged, with horns over eyes shining like water at night. The snake waved in the air, cowering, and slowly ebbed away from us. We left it in peace, in the stillness of the sand dunes, and went to bed, zipping the tent behind us.

> Phainopepla (like a black cardinal)
> Pyrroloxia (like a grey cardinal)
> Ocotilla (in bloom)
> Golden eagles

We spotted the eagles at noon the next day, two golden eagles soaring up in the sun. The sun was awash in pools of rainbow light—sundogs, plashes of iridescent color that make a halo around the sun when sunlight passes through ice crystals high in the stratosphere.

While we watched the sundogs, a roar swelled from behind the black hills, rumbling like surf rushing up a steep beach, cresting and breaking, booming. The rainbow pools shimmered and rippled, and wave after wave rolled across the iridescent surface of the sky. An answering boom echoed off the far hills, and sky waves broke from the south and pulsed back across the rainbows, intersecting the first waves, breaking the sky into fluttering diamonds of silver, pink, green.

Gradually, the rainbow waves flattened and faded. A tiny black jet flew low over the hills in oceanic silence. The eagles

circled higher and higher until they disappeared into the stratosphere, now as hard and shiny as the inside of an oyster shell. We
lay flat on our backs and watched them go.

> Golden-chain cholla
> Jumping cholla
> Coachwhip snake

We found the coachwhip snake on the dirt track as we
drove back to the campground. We stood in the center of the
road and stamped our feet until it sidled into the brush.

Back in the campground, the lady with the binoculars was
glad to see us. She had worried about us when we didn't return
to the campsite at night, she said, and then she offered lemonade,
because she thought we needed it. In fact, the only thing she had
that we needed was more time in the desert.

> Common pigeons
> Ring-billed gulls

The gulls held in the wind over the tarmac at the airport in
Phoenix. Wings fixed, they slid off streams of jet exhaust and settled feet-first onto the dry grass beside the runway. Erin wrote their
name at the bottom of the species list as our plane lifted into the air.

The Brazos River

Country-western dancing could only have been invented on the Great Plains. It takes space, lots of space. You can't dance the Texas Two-Step in the living room, even with the furniture pushed against the wall. Two-Step on out of the parlor, lengthen your stride, just keep right on moving in a straight line, *step step quick-step, step step quick-step*. Travel, that's the key—big loping strides that really cover the territory. Dance the length of Highway 95, straight down the desert side of the mountains. You could Texas Two-Step the length of Oregon in less than a week, dancing nights, sleeping under junipers during the heat of the day, step step quick-stepping all the way to California.

Frank and I are learning country-western dancing at the community college class that meets in the gym at the Presbyterian Church on Thursday nights. Our teachers are Marty, who looks like a horse in a cowboy hat, and Beth, who laughs a lot, but never makes a sound. Her mouth opens, her shoulders bob up and down, and air comes out in silent little explosions—*hah-hah-hah-hah*. Marty dances with a microphone stuck in his back pocket. He pulls it out and tells us what to do. *Okay. Everybody. Sweetheart position.* Scattered across the gym floor, couples face each other and grab hold. Start with your right foot. Everybody gotcher right foot? Aww raaht. Now.

> *She hugged me and kissed me and called me her dandy.*
> *The Trinity's muggy, the Brazos quick-sandy.*
> *I hugged her and I kissed her and called her my own.*
> *Down by the Brazos she left me alone.*

Marty and Beth dance like a Texas high school drill team—perfectly matched, flashy, with fast, precise, economical movements—and then everybody whoops and claps and Beth's shoulders jiggle up and down.

The first thing you learn in a country-western dance class is dance floor etiquette, what Marty calls *ek-it*. The basic rule is, keep moving or get the hell out of the way. Fancy stuff in the center of the circle. On the outside, movement—fast, spinning if you want, but still moving, counter-clockwise. Ladies walk backwards, men walk forwards, none of this huggy stuff because if you sag against him, ladies, he's gonna walk right over you.

The second thing you learn is that the man is in charge. Except that's not the way Marty says it. He says, *Ain't nothing worse than a lady what leads.* Gentlemen, walk her backwards.

Spin her in. Spin her out. Aww raaht. Now do it again. It's always the lady who turns, moving fast so she's back into position when the gentleman reaches for her hand, always the lady who moves her cowboy boots double-time *step step quick-step*—pointy toes, undershot heels, topstitched snakeskin, tiny things, sexy as hell. Backing away from her gentleman, watching his jaw for the tightening muscle that will tell her she's about to spin 360 degrees, a lady knows that when he pulls her toward him, he's getting ready to flick his arm and pitch her out, side-arm, like a stone skipping over a greasy river.

> *Li li li lil lil ly, give me your hand,*
> *There's many a river that waters the land.*

The men dance with their hats on, the ladies without. The way the lady ducks and tucks, if she had on a hat, it would get sent across the dance floor, first spin. Instead, the ladies wear hair. Texas-sized hair curled with a drugstore perm on pink rods, so it sort of crinkles down their backs and fizzles out halfway to their waists. And eyes. Big eyes outlined in black, and eyelashes like God used a crayon. Waists cinched with a silver-tipped belt, secured with a silver buckle the size of a salad plate. And then jeans. Ladies, you buy brand-new boot-leg forty-dollar jeans and you wash them in the Maytag, on hot. You put them on while they're still wet and wear them 'til they shrink-to-fit. The gentlemen wear Levis too, and boots with two-inch heels that boost up their rumps and bend their knees and give them a sway-backed, swaggering stance like an out-to-pasture race horse that has won its share of races. Their jeans have the look of the barn about them—faded, rump-sprung, apple-kneed, with just a bit of flare where the bottom meets the boot.

We don't really fit in, Frank and I. I think it's Frank's fault because he won't wear the *his* part of our his-n-her shirts. He wears oxford-cloth shirts instead—the only man on the dance floor with little white buttons instead of mother-of-pearl snaps. I think I've got the clothes right, but I'll always have school-teacher hair. Or maybe it's some kind of attitude problem; I don't know. We're probably too stiff, too self-conscious, the sort of people who would never make love with the horses watching. If we had horses, which we don't. Or maybe our politics are wrong. But how do they *know* we voted for gun control? How do they know that I don't let dogs in my living room, or that I get my back up when strangers call me Honey? Whatever; when the music stops and people drift toward the edges of the room, the crowd parts silently around us and rejoins, like a flock of sparrows avoiding a telephone pole.

We are immigrants to the West—emigrants from Cleveland. But we have lived in Oregon for twenty years now, and you would think that would count for something. We brag about the West as if we had created it, we plant trees and chop them down with an axe, we seek out every isolated river valley and learn the calls of the birds. But a sense of connection eludes us, and, like other immigrants, we live for those small, transcendent moments that may exist only in the imagination, when we will belong completely and perfectly to a way of life embedded in this land.

Frank and I are taking Intermediate Dance this term, having taken Beginning maybe three times before we got the courage to move on. Beginning was tough. It took its toll on us. I can feel the beat in the music and match it with my feet, but I screw up the steps. Frank's got the steps down, no problem, but he couldn't pick up a beat if it had a rope handle. So we dance an uneasy ten-step: I'm beating Frank on the back in time to the music, *ONE two THREE four, ONE two THREE four,* hoping he'll catch on, while he's muttering *BACK scuff HEEL toe HEEL step BACK scuff HEEL step,* and rolling his eyes heavenward.

Meanwhile, a lady named Mary is yelling at Dean, *Well if you weren't so damn DEAF* (she kicks him in the ankle) *maybe you wouldn't dance like you got a flat tire.* A brunette with fringe on her sleeves just walks off and leaves her man standing in the middle of the floor with his thumbs in his belt and an incredulous look on his face. So Beth goes over to dance with the gentleman, and Marty says, *Okay. Everybody. Gotcher right foot? Aww raaht. Now.*

> *The sweet Angelina runs glossy and gliding,*
> *The crooked Colorado runs even and winding,*
> *The slow San Antonio courses the plain.*
> *I never will walk by the Brazos again.*

In Intermediate, we are learning the Cowboy Waltz. This is familiar territory, so we don't have to think about the steps. Instead, we can move to the music and listen to the words of the songs, which are so hopeless that I waltz along with tears in my eyes. Frank's got his thumb hooked in my belt loop to keep me from dancing off in the wrong direction. My arm reaches around his neck and my nose presses against his sternum and I'm singing along with the music.

> *Li li li lil lil ly, give me your hand.*
> *There's many a river that waters the land.*

Real trouble doesn't come until they try to teach us Windows. Everybody is standing there, facing their partners and holding hands. First we're supposed to start moving our feet in the regular pattern. *Step step back-step/ Step step back-step.* Then, the gentleman pulls all four hands together and *Don't stop moving your feet—step step back-step* lifts them over his head. By this time, he's got the lady strung up like a side of beef and he pivots his wrists which means that she's got to spin *Two and a half times, ladies, two and a half times* or her arms will break off at the elbows. *Don't stop moving your feet.* Then the man lowers his elbows to about shoulder level and the two of them parade around with their arms framing open space that's supposed to look like cabin windows with the glass shot out. All over the gym, couples are frozen in hellish contortions and only the music moves on.

Beth snaps off the CD player, but it's too late. The cowboys and ranch wives have been transformed into ordinary people like ourselves. Not cattlemen, not ranch hands, but accountants and truck salesmen and nurses and golf pros and, yes, professors, all thinking that if they could just remember to start with their right foot, if they could just put on their ceremonial clothing and immerse themselves in the beat of the music, if they would learn to move their feet in patterns dictated by ancient practice and the rhythm of slide guitars, they would find what they are looking for.

What do we want, all of us emigrés? What is the disease that empties us from the inside and sends us out in search of who we might have been? I can only speak for myself, but there was a time when I thought that country-western dancing would be a medicine dance, a healing dance, a dance of longing for a life that is connected to a way to make a living, and a living that is rooted in the land, and a community so connected that life depends on it.

I want to be lifted off the linoleum of the Presbyterian Church and set down, dancing, on a wood plank floor, while a fiddler is sawing away, keeping time with his hips, and a little cowboy is singing the truth as he knows it. Ladies' skirts are swinging and men are stomping, and the circles are whirling, and fat, pretty babies are sleeping on featherbeds in the corners of the barn, sighing in their sleep. Outside, there are wolves and droughts and prairie fires, but we don't give a damn because the air in the barn is music and music is whiskey. Except that the janitor comes in and whispers something to Beth. She stops the music and asks if everyone would kindly remove their boots, as they are making black marks on the linoleum. Boots line up like hopeful teenagers along the edges of the dance floor.

The Jet Stream

There are two types of rivers, *continuous* rivers like the Willamette River and the John Day, and *pool-drop* rivers like the Umpqua and the North Fork of the Alsea. Continuous rivers have a generally steady slope and a uniform rate of flow. Boaters classify the Smohalla as a *22–C* river because it steadily loses 22 feet each mile, running fast and smooth, flowing across the surface of a lava bed gently tilted toward the Snake.

In pool-drop rivers, elevation changes occur quickly in steep drops separated by long areas of level water, quiet pools alternating with headlong plunges through chutes torn at the edges on broken rocks. The Rogue River in southern Oregon is a *13–PD* river, because it loses 13 feet every mile, pooling in deep eddies and dropping over Rainie Falls, Tyee Rapids, Wildcat Rapids, down step by step through Coffee Pot and the Devil's Staircase. The Snake River in Hell's Canyon is an even more dramatic *9–PD*: tons of water, deeper, swirling pools and waterfalls that smash into foamy holes and curl into rollers mined with rock.

Boaters rate the rapids in a river on a scale of one to six, where six means lethal—the chart says, *6: involves substantial hazard to life*. Few continuous rivers have rapids that rate higher than Class 2, but pool-drop rivers are almost always studded with dangerous falls. They appear on the river maps as circled *3s* and *4s* marked along the lengths of the rivers like data points on a line graph.

Each summer, for three summers, I came up out of a river canyon—the John Day, the Snake, the Rogue—and called my answering machine for news of my father, who was dying of cancer. I mapped the months when he worked steadily at his desk, filing insurance forms and giving away his field guides; I recorded the free-falls, the terrifying times when his bones crumbled and his sentences would drift and fade. The pattern became clear: a series of plateaus alternating with dramatic slides into deeper and deeper pain, the slides marked by morphine dosages plotted on a chart and by broken teeth, their edges shattered when his pain was too much to bear.

Each summer I would stand by the phone on the shady porch of a different store-cafe—usually the only store, the only cafe, at a high-plains crossroads. Beyond the shade is tawny, shimmering wheatland, cut deep by whatever river canyon I had just left. The wooden walks under my feet are smooth, the pointer on a moon-faced temperature gauge points to 90 degrees one year, 100 the next. Rufous hummingbirds buzz and swoop at the feeders. I punch in the numbers on the telephone. As I listen to the phone ring in my own home, my children push out of the store and hand me a creamsicle or a Pepsi, letting the wooden screen door slam behind them. I scan the heat-brittled notices on the bulletin boards while I listen to my voice answer the phone. *WANTED, summer range for 43 head of cattle. Farewell potluck for Pastor Martienssen. Troy-bilt rototiller with furrow attachment. Runs good.* "Please leave your message after the beep."

I would push 1, 9, and listen to my answering machine rewind. Each summer, the voices on the machine sounded more frightened. "This is Doctor Nelson." "This is Doctor Barnes." "This is Doctor Rosenberg." "Your father's failing. We think you should come."

In July, my friend Susan ran the Grand Canyon of the Colorado from Lee's Ferry to Diamond. Day after riverday, for sixteen mornings, she woke up knowing that in order to get to the landing, she would have to run Lava Falls. All the river maps confirmed what she knew also: that on the Colorado's own rapids-rating system, Lava Falls was a 10. On day sixteen, she and the oarsman scouted the rapids to look over the course they would take: head straight down the middle into the first standing wave and punch through, then aim for the exact center of the V, slide down, punch through the next wave, and then pull like hell for the left to get around a boulder and avoid the suckhole below it. What they could not see was that the first wave would flip the oarsman out of the raft, somersault him backward, and leave the boat adrift in the huge, heaving river.

Alone in the boat, Susan threw her weight onto the forward tube. The raft rode up the first wave and broke through the crest. Then suddenly the boat turned sideways to the current, straightened, and slid down the side of a lateral wave. My friend jumped for the high side, but there wasn't time. She has no memory of hitting water or of being shocked by cold, even though the air was 105 degrees and the river 47; she doesn't remember being frightened. First she was in the raft, and then she was in the water, golden light flooding down from far above. She knew she had to get air. She didn't so much bob to the surface, as the surface came down to meet her. She took a deep breath and saw a wall of water rising twenty feet in front of her. Then she was inside the wave, back in golden light. By the time the surface came down to her again, she knew she had floated through the worst of the rapid. She lifted her feet and rode out the big water.

To get to my father's house, I flew Delta. For four hours, I flew in a big DC-10 from Portland to the Greater Cincinnati Airport. Then, just across the concourse, I boarded a smaller plane, a 737, that would take me from Cincinnati to Cleveland. Cincinnati is only 250 miles from Cleveland, and as soon as the little 737 climbs up to 25,000 feet, it starts to land. The landing process goes on for forty-five minutes. The plane hovers for several miles and then drops suddenly, bouncing to the next level of sky where it slows almost to stalling, only to drop again, hold, and then drop again. It took every ounce of strength I possessed to leave Concourse D in Cincinnati and walk onto that 737, walk onto a flight I knew would terrify me, fly toward a diminished father shouting out with pain, fall through space, uncontrolled, unable to distinguish my pain from his pain, my danger from his, my dread from his despair, the smell of my own fear from his sour flesh.

What made it possible for me to fly was that I had learned to sit very still, tray table locked, seat back in an upright position, imagining myself in the bow of the drift boat on a continuous river. While other passengers saw me as a small, strange woman with her eyes squeezed shut, I was listening for a solid clank as the brass oarlocks bit into the oars and the oarsman pulled hard away from rocks. When I shifted my weight, the boat listed to starboard. The rush of air across my face was cool and smelled of algae; a waterfall must be dropping from a side canyon into the river. That would account also for the roar I heard, and the subtle shivering of the boat on rippled water. The bouncing was only the river current deflected off a cliff draped in moss and monkeyflowers. I trusted the oarsman to avoid the rocks and I trusted the drift-boat's high sloped sides to curl away the standing waves. If worst came to worst and the boat spilled me midriver, I trusted the river itself to keep me safe and carry me to a quiet pool.

Through the shadow the boat cast on the water, I could see the bottom of the river. It moved by me slow and golden, lighted by shafts of sun that seemed to come from my own eyes. The rocks were coated with a thin layer of algae marked by the trails left by tiny black snails. The designs the snails drew were intricate and beautiful, like rivers and contour lines on a topographical map. In my mind's eye, I saw a puff of dust where a crayfish had pulled back under a shale slab and the lacy skeleton of an alder leaf, dancing on end.

The current quickened and the oarsman pulled hard on his port oar to turn the boat into the landing. I felt the boat, the plane, brake, drop, then float, then hit the ground and bounce once, picking up speed on the runway, swaying from side to side, braking hard, finally slowing. With movements as familiar as breathing, I stepped over the gunnel onto the gravel bar and walked without luggage up the ramp to the taxi that would take me to my father's house.

There are two different ways to kill pain, the *anaesthetic* and the *analgesic*. The anaesthetic works at the origin of the pain, preventing damaged nerves from sending pain signals to the brain. Novocaine is an anaesthetic, and so is ether.

But analgesics work in the mind. The pain surges in through long nerve cells, but the analgesics prevent it from connecting to the pain receptors in the brain. Pain is only a feeling, and when it is not felt—when the pain signals fall away, unregistered, disconnected—it does not exist. A trembling, heightened awareness of furious activity in the mind is all, like an idea that is lost when the mind of an old man releases a thought that, unanchored, drifts away, and the man, frightened, frustrated, gropes after what is gone.

Morphine is an analgesic, and so is time, because physical pain, once it is gone, cannot be remembered, even if you try. You can focus your mind on a physical hurt you suffered years ago, or yesterday. You can rehearse detail by detail the slow-motion fall through space, the landing that shook your teeth, the sharp pain your body sent over to your mind. You can imagine it, you can describe it, but you cannot feel the pain. The pain has fallen away and cannot be retrieved.

Emotional pain is different; it stays with a person. Unexpected, unlooked-for signs bring back memories of loss that are so vivid they can't be distinguished from the original pain itself. Sights, sounds, a person's initials on the automatic dialer of the telephone, ready to dial a phone that will ring in an eternally empty house, a letter whose sender has died, leaving only the words flying air express, remembered advice from someone who loved you once—any of these can interrupt the flow of an ordinary day and drop you in a gradient so steep it will leave you breathless. There is nothing to do, then, but follow the day through to the end, bracing against the next drop.

Four

Puget Sound

It really is okay, I tell myself, that I am in a hotel room with my nineteen-year-old daughter, and a college boy I hardly know is napping on a rollaway bed beside the window, wearing only boxer shorts, whistling softly through his nose.

"Steve and I want to go down to the waterfront and have dinner together," my daughter says quietly. "Can you find something else to do?"

"Yes," I squeak. "Of course. Happy to." I can eat dinner alone, I tell myself. People do that. All the time. Some people go years without eating dinner with their daughters. Let her go.

I eat by myself in the hotel coffee shop, non-smoking, table for one. While I eat I read a book, which was Steve's idea, that boy's idea. The book is a collection of nature essays that I bought in the airport, so I am reading about ducks, and feeling conspicuous. Wood ducks lay their eggs in tree cavities or nest boxes high in trees near the water. Twenty-four hours after the baby wood ducks hatch, they are able to swim around and feed themselves, so there is no reason for them to stay home any longer. The mother duck flies down to the leaf litter below the nest box.

Then she sings a special song, the low, soft, duckish song that she sang to the babies while they hatched—*kuk kuk kuk*. When the ducklings hear that call, they climb out of the box one by one, teeter on the edge of the entrance hole, answer their mother, *pee pee pee*, and launch themselves into space. With little stubby wings, ridiculous imitations of wings, they can't fly a stroke, so they drop like stones, landing with a thump on their sternums. They pick themselves up, shrug their wing nubbins, and dotter off to the creek.

I close the book, marking my place with a salad fork. How can she sing that song, the mother duck? How can she force herself to sing the one song that will make her children leave the nest? Does she ever think, "I'll just wait a day, or two days?" Does she ever settle lower on her nest, feeling the little warm balls of babies pressing into the down on her breast, smelling the four close cedar walls of the nest box, listening to the soft songs of the river far below, letting herself believe, just for a minute, that her happiness will never end?

Late that night, long after dinner, long after I have called Frank to tell him that this is ridiculous, and he and I should have come together, just the two of us—hours and hours after that, it's the three of us again in the hotel room. An entire wall of the room is a broad curve of windows filled with the city below us—rows and rows of lights stacked to the edge of the Sound, and a fleet of freighters outlined in lights. Beyond them, darkness, and along the seam between black water and night sky, a narrow strip of sparkling light. My daughter is asleep on a bed beside me. Beyond her I can picture Steve sleeping in the damn rollaway bed.

This is absurd. Why did I let her arrange to meet him here at my conference? Why did I ever think this would work? Why couldn't they wait to get together until after final exams? I try to sleep, but I can't shut out the prickling lights of the city, and a song, maybe it's a song, will not leave my head. *If you love her, let her go.* Those aren't the right words. *If you love her, let her see.* Is this a song? Is this a sermon I am singing to myself? Have I heard this before? *If you love her, let her go, if you don't* ... I'm too tired to do anything but let the words grind in my mind, reproving me. "It's fine to know what you are supposed to do," I tell my song, "but how do you make yourself do it?" *Akrasia*, that's the Greek word: weakness of the will. I'm a philosophy professor and I'm supposed to know these things.

I'm also a mother, and I know this: For parents, child-rearing doesn't have a happy ending, so we'd better look for happiness somewhere besides at the finish. I know this, my husband knows it, and the duck must feel it somehow.

When I wake up again, I sit straight up in bed, my heart racing. The sky above the city is pink, puffy, glowing. It swells and pulses, pressing into the pointed building-tops. What time is it? I swivel around to look for the alarm clock. Four o'clock, way too early for dawn. "It's a cloud cover, moved over the city." The voice comes from the rollaway. Baritone. I stare in that direction. What is he doing awake? "Beautiful, isn't it," he says. I can't think of any answer. To me, it looks like somebody's lungs. The globs of pink sway and push, incandescent reflections of the office lights in the high city towers and the spotlights on the trawlers in the Sound. Without answering, I drop back onto the mattress and force my eyes to close. The song moves in again.

When Erin was little, she wanted to learn all the songs I knew. "Sing it with me," she would say. "Teach me to sing that song." So we would sing a song together, line by line, quietly, tentatively, then all the way through without stopping, louder, in unison. Then all the way through again at the top of our lungs.

"I think I've got it," she would say. "Stop. Stop. Let me try it by myself."

So I'd be quiet and let her try it alone. If she got stuck on a transition, I would fill in a note now and then, but mostly I just listened. "Got it," she would say. "Thanks." Then for a couple of days, I would hear her singing alone, soft, clear, true, the sound coming down the stairs or around the corner of the house. Days later, maybe weeks, hearing her sing the song, I would hum a harmony part. She would come closer and sing louder, and I would add my alto line, both of us confident, singing chords that raced and chased like two dogs running on the beach.

The first time she came home from college, she taught us new songs she had learned.

"Mind if I sing while we do the dishes?"

So she did. Turned the CD player up so loud that the bass buzzed in the woodwork, grabbed up a plate, hula-stepped across the kitchen and plunked the plate into the dish washer. Snatched the dishtowel off the rack, danced over to Frank, hooked elbows, swung him around the kitchen, all the time singing to the music at the top of her lungs.

I roll over in the hotel bed, trying not to make any noise that might signal Steve that I am awake, and I bunch the sheet up against my ears. You are being dumb, I tell myself with real anger. Shame on you. What would the world be like if people were never allowed to leave their parents? It's good to sing by yourself or with your friends. I force my mind to a stop and I lie still, cold in a queen-sized bed that smells of travellers.

When I was thirteen, I had three friends. Together we were a quartet. We sang, walking home from school. *Roses love sunshine / Violets love dew / Angels in heaven / know I love you.* Four girls in sweaters too small, skirts too short, lipstick too bright, singing in four-part harmony, walking through the neighborhoods past small houses with big front porches and tall maple

trees and sidewalks quarried from the sandstone quarry, walking slow, holding the chords practically forever.

My sisters and I made a trio. At night, when we had been
tucked into our beds and kissed good-night, when the lights
were turned out, we would lie on our backs and sing. After a
while, our mother would come to the bottom of the stairs and
call out, "Bedtime, girls." Then, ten minutes later, she would
stamp halfway up the stairs and yell, "Don't make me come up
there." We knew she would stay there and listen—demanding
silence, alert for singing—so we sang in little puffs of air with
hardly any tune, only soft simultaneous whispers, in our room
upstairs on summer nights, with car lights sweeping across the
curtains, and sometimes a car door slamming, and our mother
sitting on the stairs.

When Erin and Jonathan were little, I would tuck them
into bed, turn off the lights, half close their bedroom doors, and
then sit on the top step by the landing and sing until they went to
sleep. I remember lullabies about train whistles, and ragged
owlets, and one awful morose thing about lost children who
died in the woods, and a cowboy lullaby my mother had liked.
While I sang, I never knew if the kids were awake or asleep, so
after a few minutes I would test them by singing more and more
quietly and then fading away entirely, to sit silently, listening.

Desert silver blue beneath the pale starlight...

Long silence.

"Mom?"

"Yeah."

"Don't stop."

"Okay, but go to sleep." ... *Silver winks of light along the far skyline* ...

Good lord.

Now I sit in my meetings in the hotel in Seattle while Steve and my daughter ride the ferry across the Sound to Bremerton on a day so cold they will be huddling together in the lee of the lifeboats. I think I must have sung myself to sleep last night, silent singing, the old lullabies. We three had breakfast together this morning in the hotel coffee shop—I paid—and for all the conversation, I might as well have brought my book and read about ducks. I'm supposed to be listening to the panel of college deans, but that's futile. "It's going to be all right," I say to myself, as if I were an accident victim, dazed, sitting on an upholstered chair at the Association of American Colleges. I address myself sternly: "You will just have to learn a new part. Make it up as you go along, hum it softly, tentatively. Practice. Sing it alone until you're sure you have it right."

Klickitat Creek

IX MAGNIFICATION. Klickitat Lake was once a forest of Douglas-fir trees, one of those legendary northwest forests where huckleberries grow head-high and the trees are Goliaths draped in lungwort and dog lichen. Eighty years ago, loggers felled the forest, and judging from the height of the stumps, that was a job for giants. When a tree is thicker than a whipsaw is long, a logger has to take an axe, chop a hole as high as he can reach and wedge a springboard into the hole so that he can climb onto the board and cut the tree up high, where it narrows. They had to go fifteen feet up these trees to cut them through. To skid the big logs out, they built a road across Klickitat Creek, and when the rains came in winter, as the rains always did, the roadbed turned into a dam that made a lake out of the hollow where the forest had been.

It's a lake to this day. The old stumps stand like islands in the water, islands forested themselves now with lacy little pines and purple asters. Swallows sweep across the lake, building their nests in springboard holes rotted deep into the stumps. Honey mushrooms and red-capped lichens climb the south sides of the stumps and on the north sides, moss hangs in dreadlocks between the maidenhair ferns.

Klickitat Creek pours fresh water into a grassy marsh that filters into the lake. The creek is more a canal than a stream. Its vertical sides are troweled smooth, shaped and maintained by beavers, dredged each spring by tiny paws working into the night. The creekbed steps down over a dozen little beaver dams that drop through the valley like attic stairs. In a valley that narrow, sunlight filters in through breezy alders, so it is dancing light that beams across the yellow spathes of the skunk cabbages, dancing light that sways in the tannic water of the creek.

The pond is overhung with alders too, so when the wind blows, ripples of light reflect off the lake and move in waves through the leaves. Last year's lily pads tile the surface in some coves; in others, a thousand flowers bloom at the surface, sprouting from floating mats of vegetation. The giant trunks of felled trees crisscross the lake, forming causeways for animals. To judge from their droppings, otter and mink pad along these floating bridges, following paths that duck under the plants that grow now in the rotting surface of the great logs—willows and purple flags, blackberry vines and horsetails, buttercups, dragonfly cases.

This lake, this product of the neglect of loggers and the assiduous attention of generations of beavers, this rich play of light and life, has come to be the gathering place for thousands, truly tens of thousands, of roughskin newts, *Taricha granulosa*. When the breeze drops, when the surface of the water flashes from silver to black, it is possible to see newts floating just under the surface of the lake.

10X MAGNIFICATION. Today, March 24, Klickitat Lake's newts are almost all males. They float dully in the water with their legs dangling down, their heads drooping forward as if they were hanging by their necks, all lined up facing the shoreline, brown, rough, like rubber cigars with four stubby legs and a vertically flattened tail. Occasionally a newt pushes to the surface, burps, gasps, and sinks down again, but most just hang in the water, hang there facing land, holding in place with languid S-curving strokes of their tails.

The female newts are moving from the forests to the pond through warm, thick moss. The main migration starts in February and continues into April, as long as the warm spring rains continue. There are hundreds of them moving at any one time, thousands of them, although they are hard to see; their skinny little backs are the color of oak leaves that have spent the winter at the bottom of a creek. Their eyes are as opaque and unmoving as nuggets of gold. Their bellies are hard and swollen, their tails narrow as knife blades. Like mechanical toys, as if each front leg and the opposite back leg were two ends of the same stick, they bend their bodies this way, that way, nodding their heads back and forth as they pad through the forest. If you pick one up, she will fit in the palm of your hand, just barely, and she will keep moving her legs in the same steady rhythm, plodding onward into the air. Moving through the forests, always toward the water, sometimes walking two, three kilometers, the female newts finally tumble over the bank of the creek or push through marsh grass and fall into the pond.

When a female swims into the vision of a waiting male, he startles, then lunges toward her, swimming so fast he bumps his nose into her belly. Then he scrambles onto her back, grasps her trunk with all four legs, and pulls her tight to his stomach. She somersaults through the water, contorting her body to push at his legs with her little feet, while he rides her piggyback, denting in her sides with the power of his grip. Sometimes the male thumps her stomach with his hind legs, pulling her toward him in a set of rhythmic squeezes. When he does that, she writhes and pulls at the water. Occasionally they come up for air. But mostly they float, float for hours, for days, float hanging just under the surface in a stupefied embrace.

Sometimes another male noses into a paired couple, then climbs on top of the male, grabbing him with all four limbs and twisting violently to dislodge him. The three writhe in the water, turning upside down, slowly spinning, and soon they are joined by a fourth male, and a fifth. Males converge from upstream and down, marching along the bottom of the creek, or coming quickly, with powerful sweeps of their entire bodies, swimming with a speed that pins their front legs back against their sides. They each jump into the fray, grabbing hold of any male whose back is exposed, until forty newts have formed a squirming, turning clump as big around as a cantaloupe, with one female invisible at the center. The mating ball may last for several hours, but gradually the males sort themselves out again and wander off, searching for single females that may have just arrived at the lake.

The pairs float, the hours pass, twenty-four, thirty-six, and the male gradually increases his thumping. They sink deeper in the pond. The thumping increases. Rub, thump, thump, rub, thump, thump, thump. Finally, the female releases air from her lungs, a big bubble rises to the surface, and the pair settles—piggyback—onto the bottom of the pond. The male releases his grip and steps off the female's back. He walks forward along her length and, like a chicken laying an egg, drops a waxy capsule of sperm onto the rotting leaves. The female straddles the sperm capsule and as soon as it sticks to her cloaca, the male jumps back on—fast, and they do it all again.

A day or so later, the male drifts off and the female is finally alone. She wanders along the muddy bottom of the pond into a cove thick with elodia and the rubbery stems of waterlilies. Each time she bumps into a stick or a stem, she grasps it close with all four legs and extrudes one or two fertile eggs, laying a few each day for many days, each little egg as glassy and round as a tiny eyeball.

10,000X MAGNIFICATION. All winter and into the spring, testosterone and dihydrotestosterone gradually build up in the male newt's body. Horny pads grow on the underside of each of his fingers, and the skin on the inside surface of his arms toughens as hard and rough as a fingernail. His tail grows broader, stronger. Sperm mature in his testes. Testosterone floods into the brain areas where nerves from the nose push through, a tangle of nerve cells tight behind the eyes. The cells there are bigger, more numerous, more sensitive to the increasing amounts of vasotocin, a chemical manufactured by cells in the brain.

While a male newt floats among all the males in the pond, waiting, his brain waves are so flat that—by human standards—he would be officially brain-dead. But when a female swims into view, his brain becomes a storm of electrical activity, brain waves slamming up and down in sharp, compressed vibrations. They send messages of change through a web of messengers. Neurotransmitters: gamma-aminobutyric acid, epinephrine, serotonin, dopamine. Neuropeptides: vasotocin, gonadotropin-releasing hormone, corticotropin-releasing hormone, alpha-melanocyte-stimulating hormone, prolactin, dynorphin, beta-endorphin. Steroid hormones: testosterone, 5-alpha-dihydrotestosterone, 17 beta-estradiol, progesterone, corticosterone. So when the newt moves toward the female, he is entirely "movement toward": pure unconscious chemical desire in a tight, knobby skin.

The lengthening daylight of early spring washes the females' brains with prolactin and drives them toward the water. The female's brain sends a "walk" signal toward her spinal column, and the spinal column takes it from there, inhibiting electrical activity in the neurons in one leg while the other leg moves forward, forcing the legs to take turns, moving always toward the water, orienting to the magnetic lines of the earth, guided by an internal map of the landscape.

In the pond, in a male's embrace, squeezed and thumped and rubbed, squeezed and thumped and rubbed, hormonal changes transform the female. Once sperm are stored safely by the entrance to her cloaca, her ovaries produce gradually higher and higher levels of estrogen and progesterone, and the eggs ripen. They spurt out of her ovaries and travel down the oviducts. In the presence of vasotocin—the same hormone that stimulates the male to clasp the female—the female clasps the stalk of the water lily and pulls it tight, while an egg travels down her oviduct, past the sticky plug of sperm, and out into the water of the lake.

10X MAGNIFICATION. Each tiny bump on the newt's rough skin holds a pod of poison, tetrodotoxin, the poison of newts and puffer fish and octopi, a chemical that blocks all nerve function so that the brain stops sending signals, including the signals that stimulate the heart to pump and the lungs to inflate. There are rumors that the Coquille Indians put newts in the stew they fed to their enemies, and police reports that a Coos Bay teenager died after eating a newt on a dare. A scientist once cut himself while he was handling newts; he took careful notes as numbness climbed all the way up one arm and lodged in his chest for a whole day and night. If children pick up a newt, they should wash their hands so they don't rub poison into their eyes. Only two kinds of animals are reported to be immune to the poison's powers—the newts themselves, and a local subspecies of garter snake.

Great blue herons and cruising large-mouth bass don't often make mistakes about newts, because the newt wears a warning as bold as a highway caution sign: a bright swatch of yellow that runs the length of its belly. When the newt is touched by something hard and sharp—a raccoon's jaws, a garter snake's teeth—the newt arches its back in a spasm so violent that its tail curls around its head and the entire animal becomes a rigid curve of glowing yellow flesh. Appalling, alarming, repulsive. So the newts live long lives for amphibians, traveling long distances, floating motionless for days, and hanging in languid embrace hour after hour, year after year, so unwary, so exposed, sometimes for as many as twenty years.

The only truly dangerous enemies of the newts are Oregonians, a man standing on high land overlooking Klickitat Lake, or sometimes a man and a boy. They come in pickup trucks that sluice down the road, making a shirring sound in the wet gravel, spraying mud. Slamming the truck doors, walking in rubber boots to the edge of the bluff, talking in loud voices about "waterdogs," they shoot at the newts with twenty-twos, shooting and shooting and not stopping until they get bored. Broken newts float on the surface of the lake.

IX MAGNIFICATION. On the far bank, Frank and Miles, his graduate student, sit together on a weathered log where Klickitat Creek enters the lake. Their canoe is a few feet away, its bow wedged into the mouth of the creek. The canoe and the marsh grass shine silver in the diffused light of early morning. In red plaid shirts and green hip boots, Frank and Miles are bright spots of color moving in a litter of equipment—blue ice coolers and buckets, clear plastic boxes filled with syringes and tiny vials of hormones, aquarium nets, a long-handled collecting net, and big jars that used to hold pickles and now hold clasping pairs of newts. They bend their heads over clipboards and call out numbers and times by the clock, making notes on waterproof graph paper as they move newts from bucket to bucket. They make notes of which animals are mating and for how long. Occasionally they raise their eyes and look intently at the lake as if they hoped to see through the glistening surface.

If you ask Frank and Miles what they are doing, they will tell you that they are trying to understand the newts' behavior. By this, they mean that they are trying to figure out the astounding Rube Goldberg sequence of switches and water valves, the triggers and trip wires that connect—step by improbable step—lengthening days, for example, and the newts' prolonged embrace.

And then they will tell you that if they can explain the newts' behavior, they can, in the smallest way, begin to understand human behavior. This is because, in some respects—do you hear how carefully they speak?—we animals are all kin. Nature creates new branches on the tree of life by adding to and, less often, by taking away. The basic structures are conserved.

The same families of hormones lock onto the same receptors in newts, in night-herons, in Marilyn Monroe. There is glory in this science, and elegance, and exhilaration—the triumph of human imagination actually cracking it, actually figuring it out.

So smaller and smaller they go, into the tiny structures on the surface of cells, into submicroscopic pathways along nerves. It can be done. And it gets you to a certain point. *What is minuscule is easily dispersed,* said Lao Tzu; *what is brittle is easily split.* But getting it all back together again is the hard part. If you take anything down to its tiniest pieces, spread them out and make a diagram, and then put them all back together again, can you be sure you have the thing you started with, even if no extra pieces are lying on the ground like stray bolts from a bike? Or has it become something different, something less than the sum of its parts?

But scientists are open-minded about this. For every neuroscientist who tries to understand an animal's behavior by charting finer and finer connections, there is an evolutionary scientist who tries to understand an animal's behavior by drawing a picture on such a large scale that it stretches from the past into the future. On this account, to understand why an animal acts the way it does, you need to understand the survival value of the behavior, what scientists call *fitness;* you need to tell a plausible story about how that behavior works to the genetic advantage of the species.

The courtship of the rough-skin newt is explained as a sort of female mate-selection, the result of which is that the female ends up mating with the strongest, healthiest male. If a male is weakened by the winter, he won't be able to hold on for hours against the struggles of the female. A good hard twist, and the female is free to look further. And there is more: The longer the pair floats in the pond, the greater the chance that another male will come along and dislodge the first, and then the stronger male's genes will be preserved. In this way, all the days of struggle serve the species well.

Scientific papers fly across the Internet, another scientist is tenured, our understanding of the natural world leaps forward.

Of course, when *I* see those male newts hanging out at the edge of the lake, slouching over their bellies, belching, hanging there in a stupor, it's not reproductive fitness that comes immediately to mind; but I suppose that's beside the point, and this much I will grant: Biologists can provide a plausible explanation of the function of a behavior, the mechanism by which it has evolved, its advantages overall, the bottom line on the balance sheet. To this extent, they can answer the question: Why this, rather than something else?

But I still have a question: Why this, rather than nothing at all?

To my mind, it is the extraordinary fact that newts exist that needs explanation. What I am looking for at the pond is a sense of their purpose, what Aristotle would call their final cause, an understanding of how they fit into the grand design. The search for meaning even in the life of a newt may be a farcical effort, based on assuming without evidence that nothing is without a purpose, that nothing is pointless, without meaning. But how hard it is for human beings to give up this faith.

And so, I wonder: Maybe the newts are like Klickitat Lake itself, the accidental side-product of a purpose we never knew or have forgotten. If this is even remotely possible, then nothing could be more important than to try to uncover that purpose.

I don't know how to begin to undertake a task like that. I have no methodology. It may be that understanding of this kind will not come from analysis; instead, it may be the sort of under-standing that comes with a rush of recognition and assent: to say, *yes I understand,* the way people say *yes* sometimes when they are absorbed in music. *This speaks to me. It fits. I have been there. I un-derstand this because I am made of this; it is part of me.* Then alder leaves and beaver dams, the vaguely lilac smell of skunk cabbage, newts floating on the black surface of a pond, a field biologist— and the relationships among all these, and everything else— become a kind of language, a way to convey meaning. The only way to understand a language like this is to immerse yourself in it, give yourself over to it, suspend judgment, keep your eyes and ears open, until it starts to make sense.

The Deschutes River

In the cove at drift mile 17, the water is transparent and scarcely deep enough to float a boat. Wave action has carved the riverbed into a sculpture of the surface—lapping waves etched in sand. Close to shore, the driftboat swings on its anchor line and rocks in the wash of the swell. Frank's fly rod is hanging by its reel from a crook in the branches of a grand old juniper that arches over the cove, and his chest-waders dangle by their straps, flapping when the wind blows, kicking up their heels. Frank and I are eating lunch, sitting on the bank with our feet in the water. Jonathan, of course, is still out fishing.

Under a high sun, the surface of the water throws wavering nets of light across the riverbed, across our feet, across the sedges, the snail shells. Rows of caddisfly cases roll gently up and down in the line of bubbles where the water washes onshore. The cases are tubes made of tiny sticks and juniper scales and splinters, all stuck together every which way.

Like all insects, maybe like all of us, caddisflies move through a fixed sequence of changes in the course of their lives.

Egg to larva. Larva to pupa. Pupa to adult. To protect themselves during their larval stage, they build hard cases, putting them together from whatever they find on the bottom of the river. They pass the months backed into the tubes with only their head and legs protruding and their soft caterpillar of a body encased in the shell, like a snail. If a larva is threatened, it withdraws entirely and becomes a juniper twig blown into the river by afternoon winds, a stick in a pile of sticks. But when the danger has passed, the stick will come to life, untangle itself from the little pile of debris, and walk along the bottom of the river.

After several months, the larvae leave their tubes and encase themselves in amber capsules as glossy as buds on a willow. There, the nymphs absorb their own soft bodies and recreate themselves as adults. They rise to the surface of the river, split their skins up the back, and crawl out onto the surface tension. Transformed into dun flies with lacy wings pitched like tents, they lift their wings, pull away from the draw of the water, and blow into the air.

Jonathan casts an elk-hair caddis onto the seam where the river current meets the slack water of the cove. He has tied his fly to imitate a winged adult struggling to escape the pull of the river. That fly has a frantic, disheveled look about it—a spray of hair and a tuft of white fur. Then Jonathan moves on upriver and I lose him behind a screen of junipers. But I can still hear the river piling against his legs, and I can hear the drag on his reel, a sound like ripping cloth, that tells me he is stripping out line to lengthen his reach, heaving line across the river, stripping line off his reel, heaving it out, stripping line, until he can set his fly down on the deep water where adult fish hold.

Two years ago, Jonathan would have been wading with me in the cove, turning rocks, picking up caddisfly cases. Last year he fished the edges of the river, always in Frank's sight, or mine. This year, he moves entirely beyond our view. His jacket rests on the bank beside me, along with his fly box, an aluminum box small enough to fit in a boy's hand. I open it like a storybook and find neat rows of hand-tied trout flies—dry flies, nymphs, streamers, muddlers—each tucked under its own little clip, each a tuft of feathers and a pellet of deer hair and maybe some tinsel or flashabou tied with his daydreams to the tiny hook. I hang the jacket in the juniper, away from the dampness, and close the little box of flies.

Drift mile 19. Skookum Creek. Frank drops anchor and the boat swings around next to a gravel bar. Jonathan steps out and walks upstream. Soon his line sails out over the river. The line curls back on itself and sails forward. Another curl behind, another forward, and his line leaps out, pauses, and drops onto the water.

When Jonathan wades back to the boat, he is cradling a gleaming rainbow trout in his hands. Golden light with a glow of pink, quicksilver light playing through the colors, so beautiful, so fluid, the fish may be nothing more than refracted light, the idea of a rainbow trout. Jonathan lowers the trout into the water. When I look again, it is gone.

Jonathan ties all his own trout flies, and many of his father's. An entire quadrant of his bedroom is taken up with a fly-tying table that spills over with animal parts and pill boxes and spools of thread on spindles. A rooster's neck, a pheasant's tail, a rabbit's ear, a calf's tail. Elk hair, bear hair, grouse breast, turkey quills. Brass hooks, gold thread, black thread, silver beads—materials tucked into the compartments of a complicated wooden suitcase and overflowing from shoeboxes that smell of mothballs and rabbit skins. The finished flies are on the floor, neatly organized in plastic boxes.

It's hard to watch Jonathan create trout flies, because the parts are so small, and his hands move so quickly, and maybe because the truth of the tying, the sense of the insect, is in the private place his mind wanders while he winds the threads, the expeditions he imagines, the swiftness of the current and the boldness of the strike. But even the visible acts of creation are magical—a spontaneous generation.

Jonathan holds a little #12 hook in the jaws of a vise and squeezes down the handle to keep it steady. With two quick turns, he attaches a thread to the hook. His left hand holds a golden cord against the shank while his right hand whips the thread around to hold it in place. He pulls a tuft of angora goat fur from a plastic bag and spins it with his fingers onto the thread, then whips the furry thread around the hook to make a body. The gold cord he winds diagonally up the body, pulling a bit so that the tension dents the body into segments. He makes the insect's head from a peacock hurl, winding the iridescent barb around and around into a clump. With scissors that he wears on his fingers, he cuts fibers from a pheasant's tail and ties them in to look like legs. Standing behind Jonathan, I can see his hair, which spirals from a single point like a nebula, and the long, flat plane of his cheek. He cuts the thread, touches the end with a probe dabbed in glue, releases the hook, and holds in his hand the tiny imitation of a caddis fly larva.

The open palm extended to a parent. Look. Look what I have found. Look what I have made. Look what I have made of myself.

Trout flies are chimeras, insects built of feathers and fur, and this may be what draws Jonathan to them. As long ago as I can

remember, he has drawn pictures of fabulous composite creatures, creating them from segments of animals that must float as
freely in his imagination as foam on water. When he was little,
he stood on a chair to draw, the same chair where he now sits to
tie feathers to fur. He held his crayon in his right fist like a
dagger and on big sheets of newsprint carved the shapes of
slouching, starving panther-lizards, bears that spit flames, a
saber-toothed, salamander-footed python with hair standing up
on the back of its neck and its eyes shooting sparks, a flock of
stilt-legged birds laying purple eggs in mid-flight, a tyran-
nosauric Trojan Horse with an entire army dead at its feet.

When Jonny was a toddler, his favorite toy was an aquarium
net. He sat on his heels at the edge of every river and poked end-
lessly at the pebbles and floating weeds, turning rocks. When he
found something that interested him, he carried it in his open
hand to his father to learn its name—periwinkle, caddis case,
agate—and then, satisfied, he carried it back and dumped it in
the river. This same child, a tall teenager, stands now in swift,
knee-deep water while the current raises waves against his
calves and slaps the anchored boat around. Leaning down to
hold a net in the current, he probes with a heavy boot to dislodge
the invertebrates, then catches them as they float downriver. He
dumps the contents of the net on the deck of the boat and, leaning over the pile, pushes it apart with one finger. Bits of twigs,
skeletonized leaves, a snail shell, two fish eggs, the whole pile

wriggling, alive with larval insects. The larvae in his net will tell him what the trout are eating, and so what imitation might lure them from behind the boulders.

Jonathan knows where to put the legs on his nymphs and what colors to mix in the dubbing. He senses the right pitch of the wings and how big the head should be. He knows which insects have wing pads and how to re-create the struggle in the wings of flying insects leaving the water.

He knows these things because he has paid attention. Ever since he could walk upright, he has been happiest to be the observer, unobserved. If you focused too closely on Jonny, he backed away and, as surely as a caddisfly larva, became only surface—the opaque details of a life. Better to stand beside him and look in the direction he was looking. Soon you would be drawn into his world, a place with hawks on posts and millipedes under rocks and, at the edge of the river, the smell of juniper, and tar on the railroad tracks. And then somehow you would feel closer to what is hidden, the unchanging center.

Stand now beside Jonathan at the edge of the Deschutes, and he will show you how to see through the surface of the river to trout holding behind rocks. He will snatch insects out of the air like a magician grabs nickels and open his hand to show you a golden stone fly as big as your thumb. But he will not tell you what he hopes and fears, or what he thinks about before he goes to sleep.

At drift mile 28, I sit in the boat and watch Jonathan watch the river, the corrugated hillsides, the flocks of blackbirds scattered and rising like ashes from a fire, the Hereford steers standing up to their knees in water. A northwest wind picks up. Cottonwoods turn silver and fall warblers tumble out, dun green. Nine miles of river so far today. Whiskey Dick. White Horse Rapids. The riffle at Two Springs Ranch. Above flat water by Red Birch Camp, swallows sail low over the river, haul up close to the wind, and cant into the air on set wings. Without taking his eyes off the river, Jonathan pulls his fly rod out of its case. He bends down to snip off the fly and tie on an elk-hair caddis. Light little insects flutter over the river. A flip of water, another, as trout rise to take flies off the surface.

The Salish River

At the swimming holes, frayed swing-ropes point down into deep pools where salmon hold, heavy with eggs, waiting for the time. Overhead, clouds sag with the weight of the water they hold, waiting, holding it in, while close under the clouds, blackbirds gather in drifting flocks, silently tipping and falling in formation. Each leaf of the vine-maple tree holds on to its branch by the strength of a single layer of cells, and from each fiery leaf, from each bare twig, hangs a single drop of water. On the bank at the turn of the river, the weight of the last blackberries bends the canes. Fishermen in their boats look for long minutes at the sky and then slowly, deliberately, pull on their slickers and stow their gear under the deck. Even the river is still, pooled up black and shiny behind the shallows.

The narrow strip of beach between the fern-wall and the water is cobbled with stones the size of apples, still damp from when the river was high. Floodwater has undercut the bank here, so that the thick roots of a big-leaf maple hang down over the beach and weave a chair that I can sit on, like the cane chair that used to be next to my father's bed. A mile of hemlock forest is at my back, and a high bank on the far side of the river closes off the rest of the world. Now that Frank has rowed away downstream, the only people I see are fishermen, and they pass quickly, without speaking, maybe without seeing me at all.

I have come to the river looking not so much for solace, as for corroboration, for evidence that I have made the best decisions, the safest bets. I'm not sure what the evidence will be, although I have a feeling it won't be conclusive. Answers come from rivers only reluctantly. You have to build understanding out of small moments, you have to cut the pieces out of a day and fit them together in a new way—like puzzle pieces, or premises.

Not long ago, I was in another state, in another time, and my sisters and I were bargaining with Reverend Hartman about the scripture for my father's funeral. My father had picked out an Old Testament verse, Ecclesiastes 3, *To everything there is a season and a time for every purpose under heaven; a time to weep and a time to laugh, a time to mourn and a time to dance.*

Rev. Hartman said, "Fine, but what about the New Testament?"

"What do you mean, what about the New Testament?" I had said.

"What about the resurrection, the life, whosoever believeth in me shall not perish?"

I glanced at my sisters for permission.

"We thought we might leave that part out."

We still laugh when we think about how annoyed Rev. Hartman was, and how hard he tried not to show it, but he should have seen this coming. If ever you're going to want the truth to be told, it's at your own funeral, and Rev. Hartman knew my father had no expectations about eternal life. He was a biologist, and biologists are more interested in plant succession than in plant longevity. That's why Ecclesiastes was perfect— rejoicing in change, acknowledging the power of the evidence that nothing lives forever, honoring the importance of that fact. "I'm willing to be surprised," my father had told Rev. Hartman, and we thought he should have been grateful to get even that much from somebody who studied bacteria. In the end, we staved off the *Born anew living hope resurrection of Jesus Christ from the dead,* kept the Ecclesiastes 3, and added a verse from Psalms: *This is the day that the Lord has made. Let us rejoice and be glad in it.*

To be honest, I don't think my father felt any need for a life after death. He was an academic, and so he was an observer, fascinated by every fact, and I imagine he expected dying to be interesting enough. And besides, dying would mark the end of a long struggle with illness, and so it might well be a happy time, bringing the same quiet pleasure that came from turning in his

grade sheets at the end of a semester. It would be a time of fulfill-
ment, of ripeness, a time to be cherished, and if there was yet more
to come, well, he could think about that when the time came.

In my father's house, surrounded by the color-coded files,
the glass jars of pond water on the windowsills, little fairy-
shrimp swimming in the sunlight, covers of scientific magazines
framed on the wall, close-up photographs he had taken—a single
cicada on a blade of grass, a dandelion seedhead, full-blown—
there, in my father's house, it was easy to be sure that life today
matters more than life everlasting. Everything in the house glo-
ried in the moment, the fact of things, everything focused on
how things are, and how wonderful. All the joy-filled facts. But
when we emptied the house, sent books off to libraries and col-
lections off to museums, boxed up the piled photographs until
nothing was left but clean squares on the carpet where furniture
used to stand, it got harder. *Let us rejoice and be glad in it* was
refuted—no, worse than that, it was absorbed by the empty
spaces in the house, and I came away not knowing what to
think.

So I have come to sit by the river because I don't know what
else to do, after having sat for so many days beside my father's
bed. I reach for my parka because the air is getting darker, the
wind cooler, the season inching toward the breaking point

when the leaves, the clouds, the salmon, can't hold on any longer. I can see the time coming, see it far downstream, see it twist the leaves upside down and send them sailing off their branches. I can see it transform the grass from gold to silver, down along the gravel beds.

The wind hits first. On the beaten water, waves of pallor wash upriver ahead of the weather. Great clouds of yellow leaves fill the air and spin upstream, and suddenly a salmon breaks, a big black, red-bellied salmon charging up out of the depths, rupturing the stillness of the river, driving through the shallows, splashing with its tail fin, raising hills of water that slide over its back. I put on my wool hat and pull up my hood. The river is white with movement, scaled with fallen leaves, the riffles sliced by the dorsel fins of the driven salmon. Making desperate charges to get over the boulders and floating backward over the gravel where the water breaks, the salmon are moving toward the spawning beds by Buck Creek.

Year after year, fish runs mark the times of a river. In the summer, little glancing fish crowd the water—the parr and the resident cutthroat, fish of light. The first September rains bring the sea-run cutthroat into upriver pools where they feed under willows vague and gray in coastal mist. But the days dry out under yellow skies, and the river drops so low that in the riffles, fishermen wearing hip boots climb out of their boats and drag them banging over rocks downstream into the next deep pool. Blocked by the same shallows, the salmon wait in salt water at the mouth of the river or in the deepest tidewater pools. Heavy driving rains return in late October, raising the river to near flood, and the salmon push hard upriver—the coho and the big chinook, and by November, the first of the winter steelhead.

The times of a river are measured by the salmon, and the stages in a salmon's life are measured by its color. Each day that passes, each mile gained up a coastal river, deepens the intensity of the color of its skin and washes out the color of its flesh. A salmon entering the river is clean and silver on the outside; the flesh inside is deep liver-dark red. A dozen miles upriver, past Tidewater, its back has turned gray and the skin on its stomach is pink. Ten miles on, up by the boat landing at Lemhi Park, the salmon is lurid—purple on its back, and on its stomach the same violent red as the vine-maples, so that when the salmon leaps and twists, dark and red and sparking gold in the steaming morning, it flares like fire doused with water. Only a week later, eight miles farther up, salmon in the bedrock riffles at the mouth of Alder Creek are sooty, spent, as black as charcoal.

A spawning salmon's fins are white and frayed, and where the skin has started to die, flesh shows through, white with fungus. Its jaws have grown long and hooked at the beak, parrot-like, black-gummed, spiny-toothed, too distorted to close properly, clamped tight, gaping still. Its flesh is so soft that a fisherman leaves deep fingerprints in its side when he picks it up. And if the fisherman were to cut it open, he would find that all the color has left its flesh and flowed into the thick skeins of eggs that glow red against the black lines of its arteries.

When the first raindrops hit the water, I move in under the boughs of a hemlock. A drop bounces up where each drop of rain falls. As more rain hits, more beads of water pop up into the air. All across the pool, drops of water ping up, ping up, ping up, raising concentric circles that do-si-do through other circles and duck into flat water. The rain drives down harder and the circling sets give way to hills and valleys, crests and troughs, until the whole river is noisy with movement, chaotic and crowded. A fisherman, drifting alone in his boat, leans over the river, throwing heavy fly line through the air, not noticing the still water that gathers in the folds of his jacket and runs over into the shadows under the floorboards of the boat.

In late afternoon, the rain dissipates into a mist that sizzles on the surface of the river, raising vague streamers of steam into the hemlock trees that line the shore. Low light washes in under the clouds and touches each ripple on the river, illuminating each drop of water suspended from every leaf and twig and cone. When damp wind slides into the valley, the forest comes alive. Hemlock boughs shake off their loads, tossing up beads of light that fall where rain had fallen before.

There is *a time to weep, and a time to laugh; a time to mourn, and a time to dance.* I examine the words in my mind: their sequence, their sweep, the sounds they make when I speak them aloud. Weep. Laugh. Mourn. Dance. I carefully fold my parka into my pack, tuck my pant-legs into my boots, and start the long walk up the river.

Five

The Clackamas River

They came and got my neighbor this morning. I knew it was going to happen today, but I didn't know they would move so fast. When I came home for lunch it was obvious she was gone. The mailbox was standing open, for one thing, and even though the dog's leash was still tied to the porch railing, the dahlias were gone from the bucket by the front door, and the roller blinds were pulled down in the living room. Lillian would never have pulled down the blinds. She always drew the curtains.

I had gone over to see Lillian before breakfast this morning. I knocked, pushed open the front door and called out, "It's just Kathy," the way I always do. But the house smelled of cardboard boxes, not toast, and the living room walls were blank except for a row of clean, white squares, each with a picture hook centered at the top. A few things no one wanted still sat on the carpet, too good to throw away—the beaver-collared coat, a stack of sauce dishes, a gardenia in bloom—and next to them, Lillian sat bolt upright in a big chair, dry-eyed, her art collection propped against her knees. She wasn't going to go, she said. She wasn't as old as they thought, she had made up her mind, and she would not be bullied by her own son. And now she's ninety miles away in the Westhills Manor, probably sitting in that same yellow chair.

So I stand now with a few neighbors who are gathered across the street, watching the empty house and talking quietly, urgently, as if the house had just burned down. I tell my neighbors I feel sorry for Lillian, and I really do, but the other person I feel sorry for is myself. What will I do without Lillian? Who will call each morning at seven to say she can't unlock her front door and will I send my husband over right away to shove on it? Who will notice my comings and goings and tell me that I work too hard? Who will tell our daughter to stand up straight? And who will measure herself against our son and say, "Is this Jonathan? He's so tall, I hardly recognize him."

Lillian called at least three times a day in the week before she left, telling me to come right over, she had something for me. So now instead of having Lillian, I have a Waterford crystal pitcher, a pair of gaudy shower thongs that she never wore because they looked like they were made for an old lady, a pen-and-ink sketch of the tree peony in her backyard, and a collection of gramophone recordings of Carmen Cavallero playing songs like "In My Merry Oldsmobile." The records are heavy black plastic with a little hole in the center and grooves on one side only—I don't know what I'll do with them. Lillian said it was "wild music," not what she likes, but maybe the young people would like it—not knowing that the young people no longer have gramophones, or phonographs, or tape decks, but CD players with quadra-sound.

She gave me an egg cookbook that her daughter had written, and one about cooking soup. I stood beside her while she sat at the desk in her library and thought about what to write on the title page: "To Kathy, from Lillian, who likes soup and Kathy." I stared hard at the ceiling to keep from crying, thinking what a rare and wonderful thing it is, to be liked the way a person likes soup.

Lillian made me feel young. "I thought you were a child," she would tell me. "I said to myself, who is that child raking up the Moores' leaves?" She made me feel competent and calm. Everything she asked me to do was easily within my abilities— opening a medicine bottle, turning on a fan, disconnecting a garden hose. Except the last thing: I was not able to tell her son she wasn't going to go. I didn't even make the phone call because I couldn't figure out what was right and what was wrong.

Lillian forced me to spend part of each day doing nothing. I would no sooner get home from work than she would call. "Can you use a banana? Well then, come over right away." Never "please" or "Is this a convenient time?"—just "Come over right away." So I would cross the street, walk up her front steps, knock, push open the front door, call out, "It's just Kathy," and wait in the front hall for what would always happen next.

Lillian closes the front door behind me, crosses through the dining room, stands in the center of the kitchen, and remembers it was bananas, yes. Bananas. So she picks up a banana and turns it over in her hands, as if it were a scientific specimen of a banana. Maybe she will put it in a bag for me. There should be a bag in that drawer. I choose a bag, but it isn't the right one. Oh, and could I use some gelatin salad. Somebody gave it to Mrs. Sartwell's sister who gave it to Lillian who never eats gelatin salad. So now, gelatin salad. What to put it in?

Lillian, do you know what I could have accomplished in the time it took you to give me a banana?—I could have emptied the dishwasher, changed Jonathan's orthodontist appointment, paid the electric bill, found Erin's soccer ball, and returned a phone call from the Dean.

Everything and nothing. So I stood in Lillian's kitchen and became part of an old woman's life for awhile, until she thrust the banana at me and said, "You better go," as if I were keeping her from her work.

Lillian kept track of our family. Each time we bumped out of the driveway pulling a boat, she came out onto her porch, stood with her hands on her hips, and said, "Now where are you going?" We would name a destination and every time, she would say, in a softer voice, "I used to go canoeing. On the Clackamas. When I was young. We had picnics on the gravel bars. There were blue herons." "Yes," we'd say, "We remember you said so." Then we pulled out and, in the rearview mirror, her hands dropped off her hips and hung at her sides, and she watched us leave, witness to her own past disappearing around the arborvitae hedge, weekend after weekend.

By now, almost all the old men in our neighborhood have died. The first to go was Mr. Jensen, whose house we occupy. Mr. Jensen of Jensen Hall, Ag Extension, Oregon State University, died sitting in a chair in our living room, when it was his living room and his daughter was at the theater. Mrs. Jensen moved into a nursing home and died there last year, after twelve long years, so now we send the mortgage checks to the daughter in San Francisco.

But most of the women are still here, all widowed of course, nursing friendships that are seventy years old, living in the gracious white houses where they raised their children, close by campus, where the sweet gum trees they planted are forty feet tall and overhang the street. They came as students to Oregon State University when it was Oregon Agricultural College, in 1922, arriving on the train from Portland with their dresses in a steamer trunk. Lillian was a tall girl with mannish hair and good posture, standing stiffly—in the old photograph on her dresser—in front of the girls' dorm. She married Walter, the dean of students. Mrs. Sartwell was a cheerleader, the same Mrs. Sartwell who doesn't get out much now, except to water the geraniums on her front porch rail. Mrs. Wilson was the president of Alpha Chi Omega. She married the chairman of the food science department and had three children and then she was president of the Faculty Wives. They all graduated from Oregon Agricultural College in the B-Curriculum (no science required).

I wish I knew what they have come to know. I wish I knew how they continue to live while layers of their lives are peeled away like successive layers of skin. How did they help each other live through the departure of all the children, leaving one by one, and then the deaths of their husbands, dying one by one, a lottery of loneliness?

When all the old people leave, young families will move into those wonderful old houses, and their children will buzz up and down on their Bigwheels, and we will give them cookies and show them pictures of our children, and we will be the old people in the neighborhood. Ultimately, my friends and I will all be widows and former things, former professors, former attorneys, former RNs, who were once married to former biology profs and political scientists. Our children will worry about us, living alone, but will phone to say that it's hard to get away when you have a young family and a new job. For now, the old people are a bulwark against our own advancing age and uselessness.

Lillian's house sold in three hours for a breathtaking amount of money, and her son never told her it was sold, even though she was still living in it, as she had lived there, for fifty-eight years. It's a young couple moving in—an accounting professor and his wife, an agronomist. The crews come tomorrow to tear out the carpet and refinish the floors.

The Smohalla River (on a Cloudy Day)

I know that flecks of foam floating with the current line tell
which way the water will push a raft. I know that curves force
a raft to the outside bank. I know that when a river goes greasy,
humps up, and slides into a duck-tail of water, a rock sits close
to the surface. But ahead of us are a thousand standing waves
and a thousand rocks. Some of the rocks aren't marked by
waves, and some of the waves don't mark a rock. I'm studying
hard while I row, knowing that if I am going to be an oarsman,
I have to learn to read the river, but how can I read this? The
faster the river moves, the more opaque it becomes, and I am
completely unnerved. I hand the oars over to Frank and move
to the stern.

At dusk, we camp under cottonwoods. When the wind
moves upcanyon at night, the shifting leaves sound like a river in
a riffle. But the river in the riffle sounds like thieves in quiet conversation, whispering, murmuring, laughing aloud and then
shushing each other, secret planning. Their laughter startles me,
and I lie in the dark with my eyes open.

In the morning, even before breakfast, I put on my tennis shoes, wade into the river, and start to move rocks. I lift stones from around my feet and stack them in a curved wall, fitting little pebbles into little holes, stuffing water-drowned leaves in the cracks, bracing sticks against the current, until I have made a break-wall that catches an eddy and subdues it in a pool around my ankles. With time, the silt settles, the water clears, and I can see through the river to bedrock.

Twenty years ago, I was a beginning graduate student in philosophy, enrolled in *Phl 546: Metaphysics and Epistemology,* embarking on the first assignment: "Choose a topic for a paper. Present your topic to the professor in a scheduled conference." What I remember most about that meeting was what I wore and what the professor said. I wore a blue cotton piqué dress that I had sewn myself. It had blue buttons up the front and a brown patent leather belt that matched my shoes. I explained to Professor Brauner that I planned to write about René Descartes, the philosopher who set out in 1641 to doubt everything that he did not know for certain and found himself forced to doubt everything, including the fact of his own existence. What had happened to Descartes, I wanted to know, what had he seen, what strange turn had his life taken, that would make him wonder— even as a thought experiment—if his belief in himself was just a bad dream? That, I thought at the time, was a very good question.

Professor Brauner looked me up and down, examined the acoustical tiles on the ceiling, absorbed the fact of my buttons, my belt, and then he said this: "There is one thing you will need to learn. And that is that philosophy is not about life. Philosophy is about ideas. Life and ideas are not the same."

I never doubted him. Shame I felt in full measure. Stupidity. But never doubt. The possibility never entered my mind that a philosophy professor might be wrong about philosophy, or about life—or that I might have had the power to make a different set of definitions. Instead, I went back to my apartment, took off the dress, changed into jeans, and remade my convictions about what counted. In the end, I wrote about Descartes' concept of clear and distinct ideas.

While I fix leaks in my little break-wall, Frank sits on shore with a cup of coffee and tells me a story about the last time he fished the Smohalla. Wading upstream along a cut-bank in water knee-deep, he was feeling his way carefully with his feet, holding onto willow roots with one hand, trying to keep his fly rod from tangling in the trees. A flash of white under the surface caught his attention. He took a step forward and had just lifted his foot for the next step when a rattlesnake reared up underwater, its mouth agape, striking again and again, the white flesh of its mouth flashing in the current, flashing up and down, up and down, up, down, rocking with the current, swollen gums peeling back from the fangs, flesh falling away from the spine, swaying slowly and lethally in the eddies, its tail wedged between rocks.

"Lucky it was dead," I say, as I stand in this quiet pool, this puddle of my own creation, as I stand there on the bedrock river bottom with snails and gravel in my shoes and my socks sagging around my ankles.

"Dead snakes are safer," Frank says, "but it's the live ones that get your blood moving." He looks at me and laughs, picks up a piece of shale, weighs its balance in his hand, and wings it across the river. It bounces once, twice, then nicks the edge of a wave, stalls, and sinks.

I majored in philosophy because I was furious at college poetry. A poem said words about one thing, but when you pushed it, it turned out to be about something else. But if you thought it was about something else, then a professor would tell you, no, the poem was about what it named. Deliver me from ambiguity, I muttered at poems. Deliver me from sentences that say one thing and mean another. It was embarrassing to think a poem was about, say, building a house, when it was really about incest. I didn't have the courage to be confused or the time to be misled. If trouble is coming, tell me that trouble is coming, and don't bring on a thunderstorm and waste my time. If you don't understand something, say you don't understand, instead of calling it *holy* and expecting me to figure it out, because I can't. I went looking for a discipline where the essential value was clarity. I found it in western philosophy.

"There are persons who imagine they have earthenware heads," Descartes said, talking about persons who imagine they have earthenware heads, and nothing more subtle. I was grateful for philosophy, grateful for its directness, its simplicity, its submissiveness. I learned to do philosophy the way my professors said it should be done: Everything I wrote was clear to the point of vanishing. Objective, abstract, precise, and never, ever about life.

Last year, Frank and I visited Greece to see what we could find of ancient Greek philosophy. Greece had been suffering under a tremendous drought, and the two weeks we were there broke records for high temperatures. Arson-set forest fires had destroyed many of the old pine forests in Attica; two thousand years ago, fires cleared away the olive trees that had shaded Socrates while he taught his students that the unexamined life is not worth living. At Delphi, nothing was left of the sacred spring that once flowed from a cleft in the granite mountain, the sacred spring where Apollo leaned from his mother's lap and shot an arrow into the giant Python, where Pythias screeched and tore her hair and muttered prophecies in a language that only priests could understand. There used to be a river running along the edge of Athens, and, judging from the size of the stone-lined trench of a riverbed, it must have been a big one. But there was no river in the trench when we saw it—only gray soot from the National Road, a windrow of cigarette butts, a stray dog, a beer can rolling ahead of the wind.

There's a famous story in philosophy about how Descartes, trying to doubt everything, soon came to the realization that some ideas cannot be imagined to be false and concluded that "all things which I perceive very clearly and very distinctly are true." On the foundation of these clear and distinct ideas, Descartes rebuilt his beliefs and Western European philosophers headed out in search of clarity.

Only after twenty years is it gradually dawning on me how steep a price those philosophers paid for clarity and for the certainty that is supposed to come with it. First to go was the philosopher as a person. By writing always about ideas, never about themselves, authors transformed themselves into disembodied authorities who had no past, no future, no reason for wondering—authors whose own hopes and fears were so submerged that they could only sway slowly in the margins like ghosts. And then the range of possible subjects narrowed: the easiest things to write clearly about are the simplest, and nothing in real life is simple. So the philosophers I met in graduate school wrote about pure, slick-surfaced ideas like truth and consistency, but not about home. Not about landscape or work. "I will consider myself as having no hands, no eyes, no flesh, no blood, nor any senses," and no one will ever "be able to mislead me in anything," Descartes said, So it's no wonder that "the sky, the air, the earth, colors, shapes, sounds" faded from the philosophical discussions in my graduate seminars. Many of the ideas that remained were reduced by isolation—unrecognizable, fractured from the lives of real people. In the end, they didn't matter.

Early in the morning of the third day's camp, I wade into the shallow bay where the rafts are moored and sit down on a log that has fallen across the water and grown up with rushes and iris. I sit looking at my bare feet veined with light through water the color of whiskey, feet that look as though they have been dead for many days. A water strider rests on the surface—a little "X" of thread-like legs. It casts the oddest shadow—four black spots, each rimmed in amber light. I can smell the rapids. Part cold water, part mildewed life jacket, part juniper, the smell is clear and sharp and purifying. I watch the sun move slowly down the ramparts, cross the river, light the length of the trees, and then I feel it climb down my body—head, shoulders, back. The progressive gladness of morning.

Water is an agent of distortion and change, forcing a person to see things in new ways. Each turn of the river opens out a new landscape, something no one has ever seen before and will never see again. The landscape reveals itself in glimpses. The river hides itself in motion. It holds layers of meaning, and so it adds mystery to the landscape, a sense of complexity and risk, a sense that the important facts are hidden from view.

The word *clarity* has two meanings, one ancient and the other modern. The Latin word *clarus* meant clear sounding, ringing out, "clear as a bell;" so in the ancient world, "clear" came to mean lustrous, splendid, radiating light. The moon has this kind of clarity when it's full, and so do signal fires and snow and trumpets. But that usage is obsolete. Now "clear" means transparent, free of dimness or blurring that can obscure vision, free of confusion or doubt that can cloud thought.

For twenty years, I thought that the modern kind of clarity was all there was, that what I should be looking for was sharp-edged, single-bladed truth, that anything I couldn't understand precisely was not worth understanding—in fact, may not exist to a rational mind. I am beginning to see that this was a failure in courage. I am beginning to understand that the world is much more interesting than this, that I don't always need to know where I am, that ambiguity swells with possibilities, that possibility is ambiguous, that I miss out on the real chance when I pile rocks at the edge of a river to trap an eddy where the water will stop and come clear while the rest of the river pushes by, boiling, spitting spray, eddying upstream.

I want to be able to see *clearly* in both senses of the word. To see clearly in the modern sense: to stop a moment, stock still, and to see through the moment to the landscape as it is, unobstructed, undimmed, each edge sharp, each surface brightly colored, each detail defined, separate, certain, fixed in time and place. These are visions to cherish, like gemstones. But also, every once in awhile, to see a landscape with ancient clarity: to see a river fluttering, gleaming with light that moves through time and space, filtered through my own mind, connected to my life and to what came before and to what will come next, infused with meaning, living, luminous, dangerous, lighted from within.

Alamo Canyon Creek

We found the first rattlesnake no more than a hundred yards from the foot of Alamo Canyon. At this point, the Ajo Mountains form two parallel ridgelines, tending north and south. Alamo Canyon breaches the first ridge and gives access in both directions to the valley between them, a valley clogged with saguaro, palo verde, yellow mounds of coreopsis, mesquite with thorns two inches long, and huge blocks of volcanic rock, black rock heaped against a bleached sky. It was quiet in the canyon, and hot, the kind of heat that makes a person's face glow and his eyes narrow. Frank and I had picked our way down one side of an arroyo, across the wash at the bottom, and then up over broken basalt on the other side on a trail barely wide enough for a pair of hiking boots. The trail cut between rock ledges and brushed past thickets, detoured around slab-sided prickly pears and dropped over boulders, forcing us to walk altogether too close to vegetation that often hides snakes.

Sure enough, a western diamondback rested in leaf litter under a mesquite that brushed across the trail. It was coiled as perfectly as a Zuni pot, with its neck poking out the middle like a lily and its broad head resting on the top coil. The snake lay in easy striking range of the trail and could have picked off hikers one after another if it had chosen to, but apparently it had not. Instead, it sat quiet and cool, shining as if it were freshly waxed, while we inspected it from a decent distance.

After the first snake, we hiked on full alert, peeking under the bushes, then scanning the trail, then inspecting the margins, flinching whenever a cicada rasped in branches overhanging the track or a lizard dodged from rock to rock like a gunfighter. Even so, we didn't see the second snake until I stepped just off the trail to get a better look at the rock foundation of a cowboy's line cabin, if that's what it was—I never did get a closer look. A rattlesnake set off an alarm buzz and I whooped, leaped away, and froze in place, all before the thought of a snake had time to cross my mind.

I used to worry quite a lot about whether I would recognize the sound of a rattlesnake in time to take evasive measures. Would I stand there wondering *Is that a cicada, or is that a rattlesnake?* until the rattlesnake decided I wasn't going to get the message and nailed me? *Is the sound more like a rattle?* I would ask people, *or more like a door-buzzer?* Useless questions, all of them. The sound of that snake had not entered my consciousness before I was gone; some part of my brain recognized the rattlesnake, and that was all it took.

From a safe distance, I watched the snake raise its heavy body and sway like a cobra, darting its head around branches to get a clear shot at me, or maybe, as Frank said, to taste the air on its tongue. The snake was as thick as a man's arm, but what I had never figured out about that cliché is that anything as thick as a man's arm could also be as strong as a man's arm, and this snake broadcast strength. Rearing back to strike, drawing up to full height as if it wanted to arm-wrestle me, the snake was five long feet of power and menace and fury, but mostly power. I could see his rattle sticking up in the air like a finger, vibrating so fast it blurred, making a terrible racket. He was hot, he was mad, and I was afraid—for the first time in my life, afraid of a snake, afraid in my muscles and bones. I backed away down the trail.

Finding rattlesnakes is one of those things you want to do until you succeed and then it doesn't necessarily seem like such a brilliantly conceived project. Still, it has its advantages as a hobby. Unlike birds that are up at the crack of dawn, snakes are comfortable in temperatures that feel comfortable to humans, and so they tend to lie low until the sun has warmed the rocks, and they shelter in bushes during the worst heat of the day. When you find a snake, it doesn't fly off and leave you wondering what it was; usually it sits still while you dig in your pack for a field guide, and it watches you warily while you pull out your binoculars and try to count the scales between its eyes.

For many years, our mentor in the snake-searching business was a man everyone called Doc, a herpetologist from the university down the road. Doc could always be talked into taking university students out on field trips into the high desert canyons each spring, and our family would tag along for the fun of it. His practice was to sit in a lawnchair by an oversized campfire and tell hair-raising snake stories long into the night. As a result, nobody ever saw Doc at a decent hour in the morning. We were always impatient, us converted birdwatchers, pacing around camp, checking our watches, sure that the snakes would get tired of waiting and go take a nap. But along about nine o'clock, Doc would crawl out of his tent, bolt down a cup of coffee and suggest, as if he were the first one to think of it, "Let's go find snakes." Thirty undergraduate students fanned out across the hillsides, turning stones, peering into bushes, calling Doc to come identify what they had discovered.

He knew everything. Side-blotched lizards in the sagebrush. Prairie rattlers in the desert holly. Gopher snakes everywhere. Western whip-tail lizards under the junipers. Scorpions under a log. A striped whip snake by the road. Who would ever think all those animals would be hiding on one hillside?—a hillside as full of surprises as an advent calendar, with something scaled and occasionally lethal behind every little rock.

We learned that rattlesnakes can strike a distance no greater than two-thirds their body length. That they have heat sensors in their faces and hunt mice after dark by the heat given off by little mouse bodies. That ninety percent of the people bitten by rattlesnakes in Arizona each year are drunken males. That one percent of those die. That rattlesnakes—even roadkill rattlesnakes, we learned from sharing a meal with the undergraduates—taste like chicken. That a snake will continue to strike after its head is cut off. That in the spring, if you see one snake you'll probably see several, because they are just then leaving the caverns where they winter underground, wound together in shifting, sifting, rasping skeins.

After the second snake, we moved even more carefully up the Alamo Canyon and made a wide detour around the old ranch buildings baking in the sun. Just at the point where the canyon intersects the valley between the ridges, and two sandstone rocks lean toward each other across the trail, we thought we heard water—an insistent, fluttering sound, soft as seeds sifting in a sack, a sound that might as well have been leaves falling through dry branches or wind in cottonwoods. Great loping strides brought us to the sound, and we looked down on a thousand rivulets trickling under gravel into a rock-bound basin. Has anyone ever heard water in the desert and not turned toward it, glad, shouting aloud and climbing over rocks to the source of the sound?

Like the snake's warning, the sound of the water lodged somewhere near the muscles in my back—a lovely wash of comfort and safety, the simple assurance of something good, unmediated by any thoughts. How could anyone explain an emotion in the muscles? As far as I know, the only person who comes close is Thomas Hobbes, an English philosopher dead now for four hundred years, a materialist who thought that the body is a collection of particles. The way Hobbes described it, desire is the movement of all those particles toward an object of desire, particles simultaneously surging toward what is good. Aversion is movement away, a million particles recoiling. I used to think the whole idea was silly, imagining a microscopic chorus line of dots. But I don't think it's so silly anymore. In the desert, I felt all the parts of my body leap away from that snake, every molecule, every nerve fiber in full retreat from the slam of intracellular fear. And the joyous movement toward water was prickly, every cell responding. I dropped my pack on the gravel and knelt over the water.

Where Alamo Canyon Creek pools up under a boulder, the water carried clouds of brilliant, bilious green algae, dotted with—of all things in the desert—polliwogs, the young of the spadefoot toad. You wouldn't think an amphibian could survive a desert summer when all surface water evaporates and temperatures reach 125 degrees. You'd think that all you'd find in the way of desert amphibians would be little dried up toadskins, tough and dark under the desert sun. But the spadefoot toads show up in wet years, digging themselves out of the sand.

Thunder brings them out. The rolling vibrations of sand pounded by thunder, the echoing pap pap pap of hard rain, the low frequency waves of the heaviest storms, startle them to life after a long time of torpor, and they start to dig. Toad after toad, they pop out of the sand and hop to water for their first drink of the year, or two years, or three. They lower their hindquarters into cool, ozone-soaked water and suck up moisture through their skin.

Frank is the one who has taught me about toads, spinning long, biological tales of sex and violence. The toads mate piggy-back and lay eggs in the desert pools, the eggs hatch into polliwogs, and then the race is on to mature into toads before the sun dries up the pool. In this race, there are no scruples, only winners and desiccated polliwog corpses, little dried-up commas stuck in the crust at the edge of the rock basin. To win, some polliwogs turn to cannibalism. Those who eat the most protein from brine shrimp grow a hard, beak-like tooth on their upper lip. They turn the tooth against their egg-mates, eating each other's tails, tearing off bites of skin, growing at phenomenal rates off the rich flesh of their kin, hopping out of the basin as the last water steams out of the algae.

I took off my boots and socks and arranged them in the sun. Then I dropped down onto the rock and stretched my feet over the water, leaning back into the shade of the sandstone slab, scattering polliwogs that soon regrouped in the shadow of my right foot. I was feeling most satisfied by the progress of this hike. We had set out to encounter the desert, to find the animals that bring the landscape alive, or maybe to find the animals that enliven us to the landscape by raising our heartbeats and focussing our attention. Encounters with animals are a gold mine of interest because the more you learn about an animal, the more improbable it seems, and you realize suddenly that there is more than one way to skin a cat, more than one way to live a life. The characteristics possessed by humans—five senses, daylight vision, a certain moral compunction about eating relatives, a daily schedule, and nowhere near enough time to wait for a good storm—are facts that restrict the way humans live. But they do not necessarily impose any such limits on an animal's life. Rattlesnakes can see pictures drawn by heat. Toads can go underground without any more fuss than going to the grocery store and live a life that would be, for a human, the most exquisite, brutalizing torture, buried for years, deprived of every sense but touch.

Observing animals, you also learn that some of the characteristics that enrich human lives—self-righteousness, reflective thought, empathy, planning to a purpose—are utterly absent from the experiences of the animals, or at least that's what I am assuming from what I know about the structure of their brains. Snakes and toads have a primitive brain, hard-wired for fear and the detection of prey, for sex, hunger, and thirst. We primates have that brain too, and on top of that, the layered accretions of the cerebral cortex. So I come the closest to thinking like a snake, to seeing the world through the brain of a toad, when my body reacts to a stimulus with terror or elation and leaves my conscious mind out of the process.

I study the issue: What was I thinking about when I was thinking with my reptilian precursor of a brain, when I was frightened by the snake and yowled, and leaped back, and froze? In the effort to remember, I reconstruct the sequence of events in my mind, and at the point where the snake buzzed, I find... nothing. Nothing. No visual images. I don't even find a memory of the first sound of the rattle. No memory of jumping. A step off the trail toward the adobe bricks, a blank space (maybe a dark space), and then I'm ten feet behind where I was before, my shout echoing against the rock, my mind ablaze with interest in a snake that is still ripping off a buzz to terrify the world.

So maybe I know what it is like to think like a snake, because I know what it is like to think not at all—to act with no memory, with no decision, with no awareness, to do the appropriate thing at the appropriate time and nothing more. In a vacuum, in unawareness extended through time, a toad may live out its life, eat its sisters, absorb its tail, lay its eggs, hop out of the pond, and in increasing heat, paddle backwards into the sand, one foot down, two feet, and wait for the storms.

Or maybe I don't know what it's like to think like a snake. Maybe snakes and toads feel emotions that I can't even imagine. Why should I think that the range of human emotions has exhausted all the possibilities? Maybe the sensation that washes over a toad when he eases his rump into a desert pool is in a category of feeling entirely unknown to me.

Or maybe what the snakes and toads lack is not emotion, but consciousness. Maybe cresting waves of anger and fear and pleasure pass over a toad, unfelt—just as a human being, sleeping, may not hear a wind that slides through the canyon, bending back the branches.

Sometimes, in a desert landscape, a landscape without consciousness, emptier of intellect than any other landscape I have ever seen, I think I can feel emotion lying like heat on the surface of the sand and seeping into cracks between boulders. There is joy in the wind that blows through the spines of the saguaro, and fear in bare rocks. Anger sits waiting under stones. Exhilaration pools in the low places, the dry river beds, the cracked arroyos, and is sucked by low pressure ridges up into storm clouds that blow east toward the Alamo Canyon.

The Maclaren River

It was eleven o'clock at night in Alaska, toward the end of June. The sun was in my eyes and the sky was an enormous bronze globe that arched hard and high above my head. Inside the globe, thunderheads rose thousands of feet from black cloud-floors, and a hundred lakes shone like windows in the dark tundra plain. I was standing on the Maclaren Summit, looking down on a river cast in gold, melted and poured through the broad black gravel wash. In a hanging valley on the absolute edge of the earth, the Maclaren Glacier gleamed with amber light and fell off a cliff. The vast, glowing space between the glacier and where I stood was empty and silent, except for the sharp peeting call of a bird I could not recognize and the tiny ticking of mosquitoes against my parka hood, a sound I thought at first was rain.

I had to hold myself together because the clarity of the air made me buoyant, and if I hadn't been careful, my arms would have risen to the sky in exultation, and all the air would have left my lungs, and maybe I would have caught my breath and bounded toward the beauty like a lonesome dog, wagging its whole backside with recognition and joy. But the truth is, I'm always a little bit careful, holding something in reserve, remembering from college psychology that hypersensitivity to one's surroundings can be a sign of mental abnormality.

I obviously have mixed feelings about this. Sometimes, on dark February mornings in town when I can hardly force myself to get dressed for work, I worry about that warning from college psychology. But most of the time it makes me angry. I think, *fine:* If it's abnormal for a person's emotions to be tossed around by the weather, then I'll *be* abnormal. I'll cry at breakfast on the rainiest days, cry so hard I can't chew, and on the first warm day of spring I will drive all over town with the windows open, singing along with the Beach Boys. Who's to say that's not all right? Who's to say that the healthiest people aren't the ones who are open to the landscape, responsive to the weather, in tune? I think it's good to change with the seasons and resonate with atmospheric pressure, deep and dark, like a cello. I think the most pitiful person on earth is the one who wrote the textbook on normality, the poor climate-controlled soul who thinks mental health can be disconnected from the wind.

The hardest thing about my university job is having to maintain a constant 72 degree emotional state throughout the entire day. Euphoria in faculty meetings is frowned upon, and depression is despised. So my colleagues and I endure long days of heating, cooling, circulating—burning profligate amounts of energy to overcome our changing internal temperatures and the rising and setting sun, burning the very stuff of our cells, of our souls, to maintain equilibrium.

It seems to me that animals do a better job of this than humans do, even the Arctic ground squirrel on my tundra mountain, the one that darted up to my feet and stretched himself up to his full height, sniffing the air, peering at me so intently that he tipped over on his face. The ground squirrel must be perfectly tuned to the weather: manic-depressive, summer-winter. His life depends on it. While the big Arctic predators, the wolves, the grizzlies, are eating ground squirrels to fatten up for the winter, the ground squirrels eat too, frantically stuffing flowers into their mouths because the lowering winter sun will send them underground where their metabolism slows deep cool, sluggish, in close imitation of death. If I were a fat little ground squirrel, I would be grateful for winter. I wouldn't try to find a way to keep warm when the wind changed direction and the light came in low. Tight with anticipation, I would pass down a tunnel through air sharp with the smell of feces in spoiling haymows of Arctic grasses, past a cache of seeds, past white worms and root-threads, down into darkness, into cold perfect air that holds no light and no scent at all.

But somehow it's supposed to be different for humans. A person who is alternately calm and agitated, shifting moods the way the moon shifts phases, is (by definition) a lunatic, *fr. L. luna,* moon. By this measure, fertile women are lunatics, their moods ebbing and flowing with the phases of the moon. And what about men who move to the rhythms of the world? What about John Muir swaying wildly in the top of a tall spruce on a ridge in the high Sierras, whipped by storm winds, "rocking and swirling in wild ecstasy" while members of the civilized world, down in the little valley towns, closed their shutters against the wind. What about Sigurd Olson, a young man in an Alberta night, flying on ice skates down a broad, frozen river, while northern lights danced a tarantella in the glowing sky? "Suddenly I grew conscious of the reflections from the ice itself. . . I was skating through a sea of changing color caught between the streamers above and below. At that moment I was part of the aurora, part of its light and of the great curtain that trembled above me." Are these men lunatics too, because they are stirred by something outside themselves? If so, is that a defect? Or in some way that we don't understand, do human lives, too, depend on their ability to respond to the natural world?

On the morning after I stood on the summit, I tried to follow the path of the Maclaren River as it filtered down through the muskeg into the taiga, the land of "little sticks," where the black spruce are skinny and wretched, roughly bearded with black threads of lichen. Permafrost at ground level, shallow, poor acid soil, frigid winter—there's really not much point in such a forest, or much hope, and the trees grow as if they know this. All that morning, I pushed my way along moose trails in the taiga, through spruce and willows mostly, and wildflowers: fields of monkshood and spires of bright pink fireweed, lichen in gray splatters. Everything was wet, and my rainpants made a slip, slip noise when I walked.

There were grayling in the Maclaren, and Dolly Varden trout, and red, rotting salmon, and so I knew there would be bears—barrenland grizzlies. I have seen these bears in tundra meadows, eating wildflowers the way cows eat hay, with sideways bites that hang in bouquets out the corners of their mouths. The bears are big (they could not sleep in a single bed) and blond—the most lovable-looking creatures that ever pulled women from their sleeping bags and ate them, buttocks first, the way the Alaskan newspapers described it. So when I walked alone along the sunken trails through brush higher than my head, I took steps to protect myself; I chose a handful of round stones and put them in a little tin cooking pot tied to my pack, where they rolled and clanked against the metal to scare away the bears.

I was a leper shuffling along the streets of a medieval town, ringing his bell *clang clang* as required by law, warning the townspeople that he is poisonous, contagious, covered with sores, seeping, dangerous and ashamed. More to stop the miserable clanking than because I was tired, I sat down on a rock overlooking the river. "Most people are on the world, not in it," John Muir wrote. "[They] have no conscious sympathy or relationship to anything about them—undiffused, separate, and rigidly alone like marbles of polished stone, touching but separate." I dumped my pebbles on the ground and walked back to my car.

Later that afternoon, standing at the edge of the road on the Denali Highway, I tried to find a way into the muskeg. Staying on the edge of things was unthinkable, but walking was out of the question; there was nothing to walk on. Waist-high hummocks floating on mud or pads of sodden lichen gave way without warning to deep boggy ditches. Clouds of mosquitoes swarmed around all the willow thickets, but the thickets were too stiff and thick to walk through even if I had wanted to, and beavers had cut the ends of willows into chisels, so I didn't want to fall. But how can a person keep her balance when one boot is buried in mud and the other caught on roots?

I got back in my car and drove down the gravel highway to the first roadhouse I could find—a store-cafe, a gas pump, a satellite TV dish, and a log meat-cache high on poles to keep out the bears. I bought a fat old inner tube for a dollar.

At a culvert, I launched my tube and sat down, letting my jeans and boots dangle in the water. I floated slowly down a brook barely wider than the tube. Bulrushes and thick clumps of bluebells bent over my head. I startled sparrows, and once a muskrat. Every so often I had to get off the tube and climb awkwardly down a beaver dam that spanned the little stream. But below each dam, the water was clear and gravel-bottomed, and I could see shining minnows suspended in the roots that hung down from floating islands. The brook gave out into a little lake that was perfectly round and icy cold. I floated around looking at things (reflections, a windrow of sticks peeled clean by beavers, a hummock of grass stems each topped by a tuft of white cotton), drifting all the while in the company of a single loon.

The loon is a straight-laced, pin-striped, buttoned-down, black and white bird with a marking around its neck that looks like a preacher's necktie. Its name comes from an old Scandinavian word that means *I weep*, and it's no wonder. The loon is an ancient bird, a strong diver, an efficient hunter, and surely a good citizen, floating mute on tundra ponds. But what makes the loon a hero in my eyes is that sometimes, on clear nights in the spring and late in fall, the loon lifts itself with strong wing beats to stand almost upright on the water, raises its head to the sky, and lets loose with wild, maniacal laughter that rolls across the pond and bounces, yowling and exultant, against the farther shore.

CPSIA information can be obtained
at www.ICGtesting.com
Printed in the USA
LVHW031600211119
638113LV00009B/1452/P

9 780156 004619